THE JOURNAL OF CRYPTOZOOLOGY

Vol. 4
(2016)

CFZ Press

Edited by Dr Karl Shuker
Typeset by Jonathan Downes/Sheri Myler,
Cover and Layout by SPiderKaT for CFZ Communications
Using Microsoft Word 2000, Microsoft Publisher 2000, Adobe Photoshop CS.

First published in Great Britain by CFZ Press

CFZ Press
Myrtle Cottage
Woolsery
Bideford
North Devon
EX39 5QR

ISBN: 978-1-909488-49-6

EDITORIAL

Welcome to the fourth volume of the *Journal of Cryptozoology*. Regrettably, there were far fewer submissions to this latest volume than to the previous three (hence the delay in its publication, i.e. needing to wait quite some time before sufficient reviewer-accepted submissions were amassed), despite the fact that ever since the ceasing of publication of the now-defunct International Society of Cryptozoology's journal *Cryptozoology* many years ago, there has been a steady clamour for the establishment of a new peer-reviewed scientific journal devoted to this subject. Now, with this present journal, such a publication does once more exist, but it will only continue to do so if it receives support from interested researchers. In other words, use it or lose it.

Having said that, although some submissions did not survive the peer-review process the four papers contained in this volume did, and constitute a quartet of very different but equally interesting and significant contributions to cryptozoology. Namely, an insightful examination of the original occurrence and subsequent dispersal and gradual generalisation in meaning of a famous cryptid term; a very pertinent application of DNA analysis to alleged cryptozoological samples, a practical approach that I anticipate becoming ever more important in this field; an exhaustive and exceedingly valuable checklist of identities proposed over time for what must surely be the world's premier cryptid; and the presentation of a very comprehensive, novel type proposal for water monsters.

The journal is now actively calling for submissions in relation to Volume 5. These should be emailed to me directly (at karlshuker@aol.com). Before doing so, however, all contributors *must* ensure that their manuscripts have adhered to the journal's presentation style and requirements as given in each volume's Instructions to Contributors section, and also online in the journal's website (at: http://www.journalofcryptozoology.com). Indeed, I have had to spend so much time adapting certain submissions to meet these requirements and style that in future, any submissions not adhering to them will be returned to the authors with a request to rewrite them accordingly before they are submitted to the reviewers for evaluation.

Lastly: All subscription enquiries should be addressed to the Publisher at publisher@journalofcryptozoology.com

Dr Karl P.N. Shuker
The Editor,
Journal of Cryptozoology

PUBLISHER'S NOTE

We have chosen the king cheetah as the emblem of the *Journal of Cryptozoology* as it is the perfect example of what cryptozoology *should* be concerned with. It is a particularly uncommon mutation of the cheetah which was first noted in Southern Rhodesia (Zimbabwe) in 1926, and was initially thought to be a separate species *Acinonyx rex*. Its status as a species was conclusively disproved nearly 60 years later when king cheetah cubs were born in captivity to parents with normal markings. Cryptozoology in its purest form is about the search for the truth concerning what Bernard Heuvelmans described in 1980 as 'unexpected animals'. The truth may not be as exciting, or even as newsworthy, as many pundits would like, but it is still the truth. And in the end, that is all that matters.

The *Journal of Cryptozoology* is published and funded by CFZ Press, which is owned by the Centre for Fortean Zoology. However, it is an entity completely separate from the CFZ and entirely independent of it. We have long been aware of the need for a peer-reviewed academic journal covering cryptozoology, and we feel that it is important that it is truly international in scope and independent of any pre-existing organisation or pressure group.

We think that this is a very exciting new venture, and are immensely proud to be involved as its publishers. We are sad to see that other publishing ventures have folded for lack of funding, and so CFZ Publishing Group will guarantee to maintain funding of this project for at least the first ten volumes, regardless of sales, subscriber figures, or other outside factors.

Jon Downes
(Director. CFZ Press)

www.journalofcryptozoology.com

CONTENTS

BUCKLEY'S BUNYIP

Paul Michael Donovan
Faculty of Arts and Education,
Burwood Campus,
Deakin University,
221 Burwood Highway, Burwood,
Victoria 3125, Australia

ABSTRACT
This paper examines the Bunyip mythology in its Wada Wurrung form (a term here used to distinguish the Wada Wurrung language from the Wathaurong people) as recorded by William Buckley, and how the Bunyip has transcended racial and cultural boundaries to become central to Australian culture, and a default term for other previously-independent indigenous mythologies.

INTRODUCTION
Australia's folklore has developed over two-and-a-half centuries of cultural diversity. Stories, songs, traditions, rituals, and ideologies from every corner of the world influence our folklore. Despite the attempted genocide of Australian indigenous peoples and their languages and cultures (Clark, 1990), certain aspects of their mythology and folklore have been powerful enough, interesting enough, or pertinent enough to have survived and to have been translated, adapted, or appropriated all together into the wider mainstream Australian mythology.

One such aspect of Australian mythology is the Bunyip, an umbrella term now ascribed to a variety of indigenous folklore about amphibious monsters. Bunyips were described in early colonial encounter stories as strange water-dwelling predators covered with feathers or fur. The Bunyip is described as attacking and eating humans, and was widely feared by contact-era indigenous peoples of the Kulin Nations, specifically the Djap Wurrung, Wada Wurrung, Bun Wurrung, Daungwurrung, Nguraiillamwurrung, and the Wemba Wemba and Wergaia people, and consequently by settlers. One of the most widely-reported features of encounters with the creature is that it could often be heard bellowing and groaning in close proximity to waterways and lagoons (Smyth, 1876; Barrett, 1946; Dean, 1998; Holden, 2001). Its call was reportedly a distinctive two-tone booming that resounded in the night, striking terror into all who heard it (Heuvelmans, 1965; Smith, 1995; Campbell, 2010).

Variations on Bunyip legends are ubiquitous. A diverse array of such monsters existed in contact-era indigenous Australian cultures. The Bunyip now features widely in modern Australian children's literature, and there have continued to be sightings of the beast by settlers since the 1830s. One of the earliest such sightings was reported by early colonial explorer and escaped convict William Buckley (1780-1856), originally transported to Australia from England, who lived among the Wathaurong people between 1803 and 1835. The first European to integrate and live long-term among the Wada Wurrung people, Buckley made ground-breaking anthropological progress in understanding, recording, and eventually translating Wada Wurrung mythology (Morgan, 1852).

Buckley describes the Bunyip as: "A very extraordinary amphibious animal which the natives call Bunyip...covered with feathers of a dusky grey colour...about the size of a full grown calf...the creatures only appear when the weather is very calm, and the water smooth...[the natives] had a very great dread of them, believing them to have supernatural power...as to occasion death, sickness, disease and such like misfortunes...they seldom remain long in neighbourhoods after having seen the creature." (Morgan, 1852).

It should be noted that although Buckley's story was published in 1852, it contains stories from his life between 1803 and 1835, including indigenous folklore, language, and other anecdotal history.

THE ORIGINS OF THE BUNYIP STORY

When examining the story of the Bunyip, it is important to distinguish Buckley's Bunyip from other indigenous mythologies. Non-Wathaurong stories, traditions, and reported sightings of lagoon-dwelling and amphibious mythological beasts have been reported from both indigenous and non-indigenous people down through the centuries (Dean, 1998) - unidentified creatures whose descriptions resemble a wide range of animals ranging from crocodiles (Anon., 1859), emus, platypuses, and thylacines, to diprotodonts, codfishes, seals, manatees (Anon., 1847f), and various other actual and mythological creatures. Their geographical origins range from the rainforests of Queensland to the outback of New South Wales, through Victoria and Tasmania (Barrett, 1946; Heuvelmans, 1965; Smith, 1995; Campbell, 2010).

Among the Victorian tribes, there were variations on the Bunyip myth, as well as other similar creatures with other names, later to be assimilated into the Bunyip mythology. Colin Leslie Dean describes these and other indigenous mythologies in great detail in *The Religions of the Pre-Contact Victorian Aborigines* (Dean, 1998). Dean cites W. Thomas, A. Massola, and R.B. Smyth in distinguishing local variations on the myth. According to Dean, the word 'Bunyip' appears in Wada Wurrung, Djabwurrung, Daungwurrung, and

Woi Wurrung mythologies. Dean also states: "The Bunwurrung called the Bunyip Toor-roo-dun" (Smyth cited in Dean, 1998). Ideas of the Bunyip appear on the Murray River and along the coast and swamp lands of the Western district (Stanbridge, 1861).

Dean also explores Myndye or Mindi (Thomas, 1847), a giant supernatural snake, who, according to Dean, and like the Bunyip, was ubiquitous among Port Phillip tribes over a large portion of the central and North Western district of Victoria. Dean describes Mooroobul, who lived in a deep water hole in the Moorabool River, and Tooroodun, who inhabited the mouth of Stawell's creek near the township of Tooradin, as similar amphibious mythological beasts in this area (Massola, 1968). Countless others undoubtedly exist in traditional lore elsewhere throughout Australia.

Many of these independent mythologies have been retrospectively homogenised by non-indigenous writers, and categorised as Bunyips. 'Bunyip' thereafter became an umbrella term for mythological creatures, which are independent and have no connection to the Wada Wurrung and Djab Wurrung Bunyip. The one commonality is that they almost invariably inhabit deep water.

In fact, the term 'Bunyip' originated as a Kulin term from Western and Central Victoria (Blake, 2011), and first appeared in print in July 1845, when the *Geelong Advertiser* reported a new animal discovered in the colony of Port Philip (Anon., 1845a). This article dates the use of 'Bunyip' to 1845, some seven years before Morgan's use of it on behalf of William Buckley, which verifies Dean's etymologies, indicating that the term existed in this area before Morgan published it. The article concerns a bone fragment, positively identified by independent local indigenous people as the Bunyip, and describes the creature in some detail based on sketches and verbal descriptions of it given by the identifiers. It assumes the Bunyip to be a genuine (living or extinct) animal rather than an indigenous mythology.

Ronald Gunn in 1847 attributes the origin of the Bunyip mythology to the "Port Phillip Aborigines" - a generic term for the Kulin people - authenticating William Buckley's use of it there as cited by Morgan (1852). Gunn also describes the physical and behavioural attributes of the Bunyip taxonomically, i.e. "about the size of a calf, with the head and neck of an emu, the mane and tail of a horse, tusks and flippers, laying large eggs in a nest similar to that of a platypus" (Gunn, 1847; Dean, 1998), and links the term with several other terms from neighbouring languages to describe a similar beast (Gunn, 1847).

Aldo Massola investigates an indigenous sacred site at Challicum in Djab Wurrung (Chap Wurrung) country, where a ground drawing of the Bunyip was carved annually by the local indigenous people to commemorate an event mythologised in their folklore, where an Aboriginal man and his brother were attacked by the Bunyip. When the man had been killed, his brother speared and killed the Bunyip, leaving its body on the ground. When their people came to organise the funeral of the killed man, they traced around the body of the Bunyip, heaping the land onto the silhouette, and returned annually to re-carve the figure, memorialise the event, and share the story of the Bunyip. Massola remarks that if

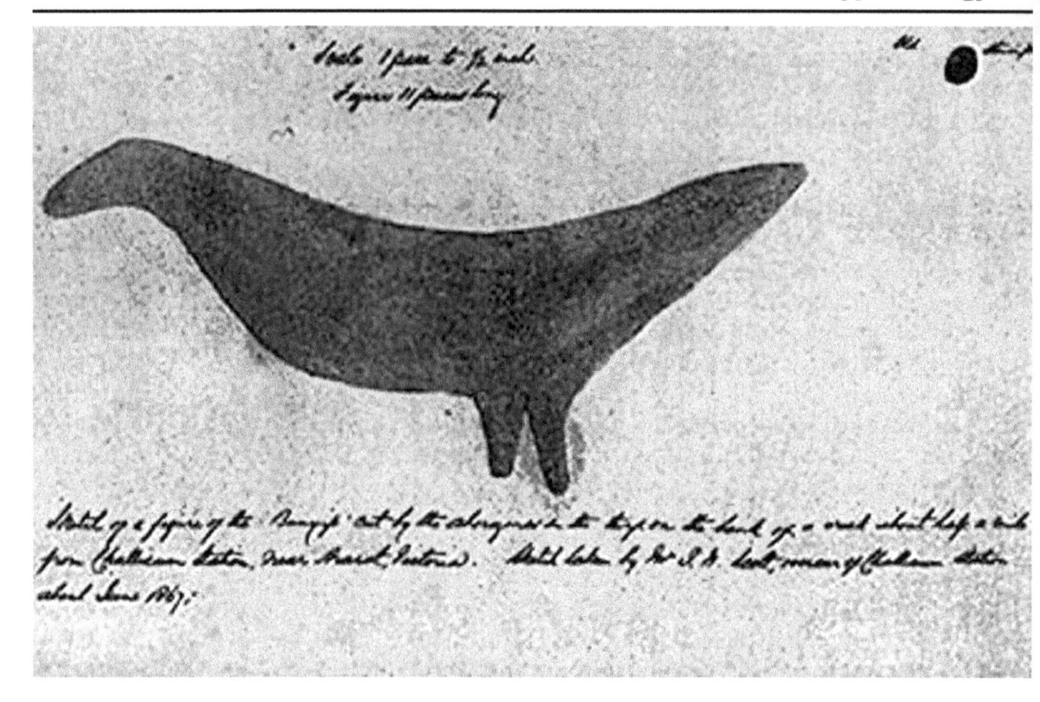

Fig. 1: Sketch made by J.W. Scott in 1867 (public domain)

the supposed head of the beast in the geoglyph is a tail, and the legs are flippers, the picture resembles a seal (Massola, 1957; Massola, 1968; Cooper, 1987). Massola explains how after the last Aborigine involved in the ritual had died, the site was fenced off for its conservation but eventually grew over and was opened for the running of cattle, leaving no trace of the geoglyph. Massola cites a map drawn in 1867 by Jim W. Scott of the Fiery Creek, including the site of the Bunyip geoglyph (Scott, 1867).

Given that the above sources agree in placing the origin of the term among the Kulin languages, and that according to Ian D. Clark these peoples were on good terms and in frequent communication in the form of inter-tribal marriages and gatherings (Clark, 1990), it is probable that oral traditions and mythologies were shared among these contiguous peoples. As the Djab Wurrung Bunyip seems to be the only one in the group with a provenance in an event (actual or perceived), it seems that this is the origin of the Bunyip, passed orally to others, potentially as a warning against an actual danger, through kinship networks and fraternisation, thus invoking fear in a creature that nobody had personally seen, but that everyone knew someone who had. Stanbridge's explanation of the stars representing the two men who fought the Bunyip (Stanbridge, 1857), one being its victim, the other its slayer, agree with Dawson's account of the two men who encountered the beast near Challicum in Djab Wurrung country (Dawson, 1881, cited in Barrett, 1946).

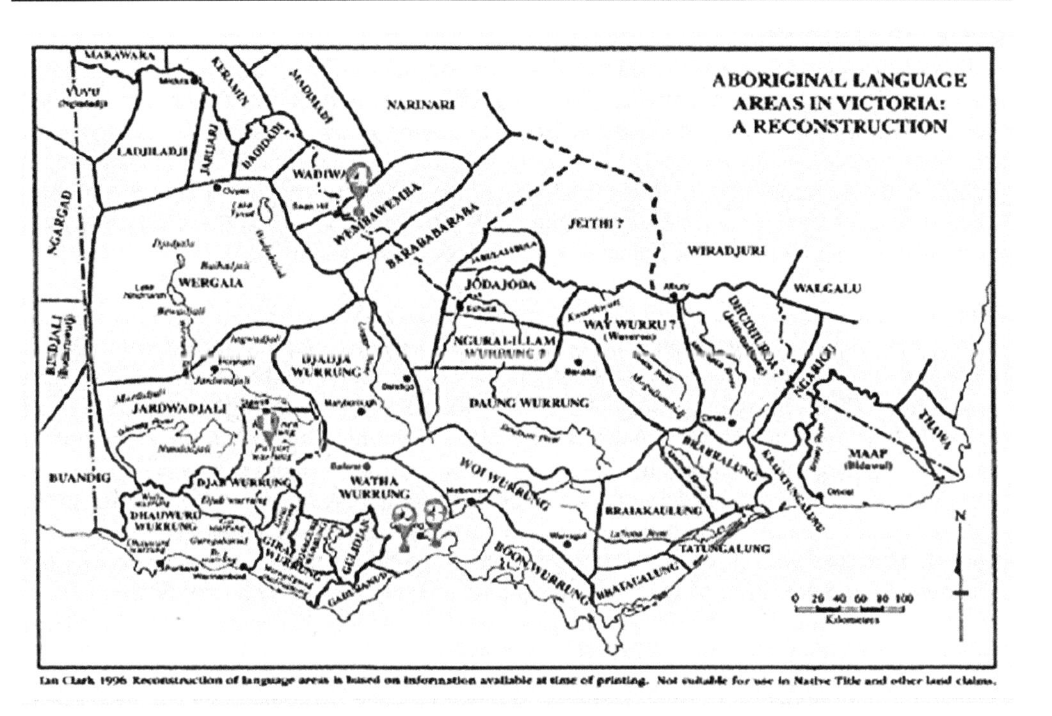

Fig. 2: A reconstruction of Aboriginal language areas in Victoria

1. Challicum, Djab Wurrung (Chap Wurrung) country, the site of the Challicum Bunyip geoglyph. Also the site of James Dawson's account of the Bunyip myth (in Barrett, 1946).
2. Lake Modewarre, Watha Wurrung (Wadda Wurrung) Country, where William Buckley first describes the Bunyip.
3. Jerringot, Watha Wurrung (Wadda Wurrung) Country, the second place where William Buckley reports the Bunyip as being frequently sighted.
4. Murray Downs, Wemba Wemba Country, the site of the Bunyip skull reported in the *Geelong Advertiser and Squatters' Advocate*, 1845, p. 2, and W.M. Edward Stanbridge's study of indigenous astrology.

Philip A. Clarke claims that 'Bunyip' originates as a Wemba Wemba or Wergaia word (Clarke, 2007). However, whereas his sources indicate that the word (and its variations – 'Panip', 'Puneep Daw', 'Wim Panip', 'Panyip', 'Bun'yipp') all exist in those languages, there is no evidence that they originated there. These languages are culturally and geographically contiguous to the Djap Wurrung country. Clarke has not made the connection with the originating myth of the Challicum Bunyip, whose ritual tradition pre-dates the Murray Downs Bunyip, and I argue, is the root of the word.

BUCKLEY'S BUNYIP

William Buckley was one of the foremost sources of ethnographic information of contact-era Wada Wurrung culture between 1803 and 1835. His collection of memoirs, entitled *The Life and Adventures of William Buckley: Thirty-two Years Amongst the Aborigines of the Then Unexplored Country Round Port Phillip*, was written by John Morgan, a journalist (Morgan, 1852). In 1803, Buckley escaped the Sullivan Bay Penal Settlement, now Sorrento, Victoria, and survived against all odds to live among the Wada Wurrung Balug people, whom he met at Indented Head (Clark, 1990; Cahir, 2001).

Being that Buckley was barely literate (Morgan, 1852), and unwilling to share his story or even speak to his colonial contemporaries who had tried to communicate with him to document his biography, causing them to think him completely stupid and/or rude (Gellibrand, 1836; Bonwick, 1863; Barrett, 1948), he entrusted Morgan, who had lived among Aborigines of various nations himself, to publish his memoirs "for mutual benefit" (Morgan, 1852). Morgan admits in his introduction to Buckley's narrative that the story was given him retrospectively, piecemeal, anecdotally, and in no particular order over the space of six months while the two stayed together in an ongoing interview process, and that Morgan had no choice but to employ some artistic licence in the arrangement and narration of Buckley's encounters. Being the only remaining first person narrative from Buckley's memory, Morgan's work is widely accepted as the historical authority on William Buckley's life and adventures.

Buckley's account of the Bunyip is brief in his memoirs. However, along with a variety of newspaper articles (Anon., 1845a,b; Anon., 1847a,b,c,d,e,f; Anon., 1859; Anon., 1892; Anon., 1932), it may well be the catalyst for this obscure pre-contact Kulin myth, exploding into the Australian zeitgeist, and thus eventually making its way into mainstream popular culture as an Australian Bigfoot or Loch Ness Monster.

The fact that Buckley cites 'Bunyip' as the Wada Wurrung word for the animal is significant. This places the Bunyip mythology in pre-contact Wada Wurrung country, and offers an attempt at a description of the beast. The sketchy nature of the description, and the fact that no Wada Wurrung people had personally seen the animal yet literally believed that it existed, indicates a second-hand knowledge of the Djap Wurrung myth from the Challicum ritual. This incidence positively identifies the mythology as local.

James Bonwick followed Morgan's work after Buckley's death, adding broader context to Buckley's life and times through ethnographic enquiry into Wathaurong histories and mythologies: "Assuming the role of a critical, unbiased historian" (Barrett, 1948), in order to provide an educational, academic history of Buckley's time with the Wada Wurrung. Bonwick makes no reference to Buckley's encounter, although, like other chroniclers of his time, he asserts that the Bunyip is an actual creature, rather than a myth.

Their belief in the fabulous Bunyip has been placed among their superstitions. This monster was said to emerge from the lakes and carry off children, and even women; but the natives never attempted to kill one. Some early stock-keepers claimed the presence of a large amphibious animal, covered with hair. But since the more comprehensive settlement of the country, the Bunyip has disappeared, to dwell with the banished ghosts and fairies of the olden times.

REPORTED BUNYIP SIGHTINGS

Following the first time that the word 'Bunyip' appeared in print, in July 1845 (Anon., 1845a), announcing a new animal discovered in the colony of Port Philip, the mythology took on a new form. Settlers were discovering a foreign and exotic environment, and were attempting to understand an alien landscape and pantheon through the lens of well entrenched Eurocentric naturalist and supernaturalist paradigms (Waldron and Townsend, 2012). Once the story of the Bunyip had entered mainstream colonial culture through the media, settlers were wary of the dangerous creature and began sighting it in their work and travels, lurking in the shadows around deep water. This became a concept with which to explain the unknown through a comfortable term (Waldron and Townsend, 2012).

One of the first incidences to validate the Bunyip mythology as an animal in colonial Victoria was the finding of a then-unidentified skull in the Murray Downs/ Murrumbidgee area. Shortly after this incident, there was a spike in reports of other Bunyip sightings throughout the New South Wales colonies (Anon., 1845a,b; Anon., 1847a,b,c,d,e,f; Anon., 1859; Anon., 1892; Anon., 1932). In 1847, the skull was shown to several aboriginal groups, who had not been in communication with one another, and all of them positively identified the skull as having belonged to a Bunyip. The skull was exhibited in the Colonial Museum of Sydney, but upon later scientific examination it proved to be the skull of a severely deformed foal (Anon., 1847d). It is this incident that gave rise to the common idea that the Bunyip myth originated here, as the word was independently used by Wemba Wemba and Wergaia people. Yet just because it did exist in their language, this does not necessarily mean that it did actually originate here. A *Geelong Advertiser* article published on 19 February stated that the local names of the beast at that point were 'Kine Pratie', 'Katenpai', and 'Tunnerbah', but compares it with the Bunyip from Corrangamite (Anon., 1847b). This also suggests that the Bunyip myth did not originate there.

Between 1845 from the "wonderful discovery of a new animal" (Anon., 1845a) and 1847 with the finding of the skull, the story had been distributed verbatim to newspapers throughout the Australian Colonies (Anon., 1845b) and a wave of sightings ensued. The latter began describing their sighted beasts in terms very different from the original nightmarish man killing wayward seal, and began describing the animal as everything from the mundane, turtles, crocodiles (Anon., 1859), cows and camels, and even a whale (Anon., 1847b,e) to the supernatural, dragons, sea serpents, kelpies, spirits, nymphs, dinosaurs, dragons, etc. (Barrett, 1946; Heuvelmans, 1965; Smith, 1995). Thereafter, different "types" or "versions" of the Bunyip occur. The later descriptions of the Bunyip as having the head of an emu, features of a crocodile, etc, all post-date these events.

THE SEAL IDENTITY

Many scholars and cryptozoologists (Anon., 1932; Fenner, 1933; Barrett, 1946; Mulvaney, 1994; Drinnon, 2012) agree that the origin of the Bunyip mythology before it was homogenised with other myths stemmed from inland indigenous communities with no knowledge of marine life encountering wayward seals that had strayed inland, thus mythologising their appearance, and passing on the experience through oral tradition. Seals can be found in inland lakes and waterways after flood tides. This theory is in keeping with the fact that the Challicum Bunyip (Cooper, 1987) resembles a seal, and that all original indigenous accounts of the Bunyip were consistent with this description prior to the 1845 press release (Anon., 1845a) in which it takes on a fantastical dimension. Astrological traditions of the Wemba Wemba people (Stanbridge, 1857; MacPherson, 1881; Christies and Bush, 2005) agree with the story of the Challicum Bunyip, offering a continuum of oral tradition from the place where the mythologised event took place to the region purported to be the term's origin.

THE BUNYIP IN POPULAR CULTURE

Once the myth reached the interstate newspapers, settlers, journalists, fishermen, farmers, and folklorists developed an awareness of the idea of a Bunyip, and were thus on the lookout for such an entity in their daily travels. Immediately, other mythologies began to be identified as types of Bunyips, Aboriginal mythologies and descriptions of crocodiles and other predators were superimposed onto an indigenous idea foreign to places such as Sydney, Hobart, Perth, or far north Queensland. Sightings of other strange and unknown creatures began to be explained as Bunyips, as this had now become a familiar image; an elusive creature within our taxonomical paradigm. Here new features began to be added to the description of the creature, until the result was part beast, part bird, part marsupial, part reptile, and part fish. In a colonial frontier whose actual animals undermined everything that the old world knew about taxonomy and fauna, where such a ridiculous idea as a platypus, a kangaroo, or a koala were commonplace, no confusion of attributes was too outrageous to be believable.

This new myth wove its way into modern Australian popular culture and folklore, and appears in 'Dot and The Kangaroo' (Young, 1977), 'The Magic Pudding' (Lindsay, 2008), and a myriad of other fictional stories (Fleming, 1939; Wagner and Brooks, 1973; Wrightson and Horder, 1973; Scott and Freestone, 1991; Svendsen *et al.*, 1994; Sibley, 1998; Herrick and Norling, 2005; Brzezinski and Tisdall, 2007; French and Whatley, 2008). It has also expanded from popular culture into science and academia, where the Bunyip has become the subject of cryptozoological study in Australia (Heuvelmans, 1965; Smith, 1995; Fooffstarr, 2009; Campbell, 2010; Tim the Yowie Man, 2013; Ghost Hunters Paranormal Dorset Team, 2013; Williamdefalco, 2014). In addition, many Australian towns have adopted the Bunyip as their civic identity, creating Bunyip-themed tourist attractions. However, local indigenous people argue that the mainstream Bunyip images are not consistent with their mythologies (Clarke, 2007; 'RR', 2013a,b).

CONCLUSIONS

After 1847, the word 'Bunyip' was understood as a pan-Aboriginal term for any unknown thing which was to be feared, and was used by settlers to describe unknown indigenous creatures, and by Aborigines to describe feral and introduced species that they did not understand as part of their environment (Anon., 1892; Dawson cited in Barrett, 1946; Taplin cited in Clarke, 2007)).

Any reference to the Bunyip that originates post-1847 and outside the original Djap Wurrung and surrounding vicinity is an independent mythology, borrowing from the name of the Djap Wurrung myth, and homogenising a variety of creatures, actual and mythological, to produce a modern, colonial mythology with its own overarching identity. This has resulted in misunderstanding, homogenisation, and narrative fragmentation, hence the confusion as to the image, appearance, and identity of the creature.

REFERENCES

- Anon. (1845a). [Untitled article]. *Geelong Advertiser and Squatters' Advocate* (Geelong), 2 July.
- Anon. (1845b). [Bunyip article]. *Sydney Morning Herald* (Sydney), 12 July.
- Anon. (1847a). [Bunyip article]. *Sydney Morning Herald* (Sydney), 9 February.
- Anon. (1847b). [Untitled article]. *Geelong Advertiser and Squatters' Advocate* (Geelong), 19 February.
- Anon. (1847c). [Bunyip article]. *Sydney Chronicle* (Sydney), 26 May.
- Anon. (1847d). [Bunyip article]. *Sydney Morning Herald* (Sydney), 7 July.
- Anon. (1847e). [Untitled article]. *Geelong Advertiser and Squatters' Advocate* (Geelong), 27 July.
- Anon. (1847f). [Untitled article]. *Bell's Life in Sydney and Sporting Reviewer* (Sydney), 13 November.
- Anon. (1859). Australian natural history. *Australian Home Companion and Band of Hope Journal* (Sydney), 19 November.
- Anon. (1892). The camel in Australia. *Queenslander* (Brisbane), 31 December.
- Anon. (1932). Queer beasts of the bush. *Argus* (Melbourne), 13 August.
- Barrett, C. (1946). *The Bunyip and Other Mythical monsters and Legends.* The Mail Newspapers Ltd (Adelaide).
- Barrett, C. (1948) *White Blackfellows: The Strange Adventures of Europeans Who Lived Among Savages.* Hallcraft Publishing (Melbourne).
- Blake, B.J. (2011). *Dialects of Western Kulin, Western Victoria: Yartwatjali, Tjapwurrung, Djadjawurrung, La Trobe University* (Melbourne).
- Bonwick, J. (1863). *The Wild White Man and the Blacks of Victoria* (2nd edit.). Fergusson and Moore (Melbourne).
- Brzezinski, M. and Tisdall, B. (2007). *The Burrawang Bunyip From the Bottomless Swamp.* Aussie Oi Oi Oi (Tallebudgera Valley).

- Cahir, D.A. ('Fred'). (2001). *The Wathawurrung People's Encounters With Outside Forces 1797-1849: A History of Conciliation and Conflict.* David ('Fred') Cahir (Mt. Helen).
- Campbell, M. (2010). Horror on home turf. *Age* (Melbourne), 19 June.
- Christies, P. and Bush, M. (2005) *Stories in the Stars: The Night Sky of the Boorong People.* Museum Victoria (Melbourne).
- Clark, I.D. (1990). Aboriginal Languages and Clans: An Historical Atlas of Western and Central Victoria. *Monash Publications in Geography*, No. 37: 227-229.
- Clarke, P.A. (2007): Indigenous spirit and ghost folklore of "settled" Australia. *Folklore*, 118(2): 141-161.
- Cooper, D.E. (1987). *The Challicum Sketch Book, 1842-53, and Supplementary Paintings by Duncan Elphinstone Cooper: Reproduced from the Originals Held by the National Library of Australia, Books 1842-1853.* National Library Australia (Canberra).
- Dean, C.L. (1998). *The Religions of the Pre-Contact Victorian Aborigines.* Gamahucher Press (Geelong).
- Drinnon, D.A. (2012). Dead bunyip preserved in outline. Frontiers of Zoology, http://frontiersofzoology.blogspot.com.au/2012/02/dead-Bunyip-preserved-in-outline.html 27 February.
- Fenner, C. (1933). *Bunyips and Billabongs*. Angus and Robertson (Sydney).
- Fleming, W.M. (1939). *Bunyip Says So : A Tale of the Australian Bush.* New Century Press (Sydney).
- Fooffstarr (2009). The bunyip mystery explained? Above Top Secret, http://www.abovetopsecret.com/forum/thread437207/pg1 15 February.
- French, J. and Whatley, B. (2008). *Emily and the Big Bad Bunyip.* HarperCollins (Pymble).
- Gellibrand, J.T. (1836). *Memorandum of a Trip to Port Phillip*. Original unpublished manuscript. Held at the State Library Victoria, Australia.
- Ghost Hunters Paranormal Dorset Team (2013). Bunyip, What is a Bunyip?: Cryptid. YouTube, https://www.youtube.com/watch?v=ounXtaEWn_k video uploaded 14 September 14 [but no longer accessible].
- Gunn, R.C. (1847). On the "bunyip" of Australia Felix. *Tasmanian Journal of Natural Science*, 3 (January): 147.
- Herrick, S. and Norling, B. (2005). *Naked Bunyip Dancing*. Allen and Unwin (Crows Nest).
- Heuvelmans, B. (1965.) *On the Track of Unknown Animals.* Hill and Wang (New York).
- Holden, R. (2001). *Bunyips: Australia's Folklore of Fear*. National Library of Australia (Canberra).
- Lindsay, N. (2008). *The Magic Pudding: The Adventures of Bunyip Bluegum*. Angus and Robertson (Pymble).
- MacPherson, P. (1881). Astronomy of the Australian Aborigines. *Journal and Proceedings of the Royal Society of New South Wales*, 15: 71-80.
- Massola, A. (1957). The Challicum bunyip. *Victorian Naturalist*, 74: 76-83.

- Massola, A. (1968). *Bunjil's Cave: Myths, Legends and Superstitions of the Aborigines of South-East Australia.* Lansdowne Press (Melbourne).
- Morgan, J. (1852). *The Life and Adventures of William Buckley: Thirty-two Years Amongst the Aborigines of the Then Unexplored Country Round Port Phillip.* Archibald MacDougal (Hobart); 2002 edit. edited by Tim Flannery. The Text Publishing Co. (Melbourne).
- Mulvaney, J. (1994). The Namoi bunyip. *Australian Aboriginal Studies*, No. 1: 36-38.
- 'RR' (2013a). The Murray Bridge bunyip. Centre for Fortean Zoology Australia, http://www.cfzaustralia.com/2013/04/the-murray-bridge-Bunyip.html 13 April.
- 'RR' (2013b). Bunyip billabong retains air of mystery. Centre for Fortean Zoology Australia, http://www.cfzaustralia.com/2013/11/bunyip-billabong-retains-air-of-mystery.html 21 November.
- Scott, J.W. (1867). Tinted drawings. Held at the Museum of Victoria, Melbourne.
- Scott, M. and Freestone, P. (1991). *The Bunyip in the Billycan.* Jam Roll Press (Nundah).
- Sibley, I. (1998). *The Bilby and the Bunyip: An Easter Tale.* Lothian Books (Port Melbourne).
- Smith, M. (1995). *Bunyips and Bigfoots: In Search of Australia's Mystery Animals.* Millennium Books (Alexandria).
- Smyth, R.B. (1876). *The Aborigines of Victoria.* George Robertson (Melbourne).
- Stanbridge, W.M.E. (1857) On the astronomy and mythology of the Aborigines of Victoria. *Proceedings of the Philosophical Institute of Victoria, Transactions*, 2: 137–140.
- Stanbridge, W.M.E. (1861) Some particulars of the general characteristices of astronomy and mythology of the tribes of the central part of Victoria southern Australia. *Ethnological Society of London (New Series)*, 1: 301-302.
- Svendsen, M.N., *et al.* (1994). *The Bunyip and the Night.* Jam Roll Press/University of Queensland Press (St Lucia).
- Thomas, W. (1847). Brief account of the Aborigines of Australia Felix. *In:* Brides, T.F. (Ed.) (1969), *Letters From Victorian Pioneers.* Heinemann (Melbourne).
- Tim the Yowie Man (2013). The bunyip hunt. *Sydney Morning Herald* (Sydney), http://www.smh.com.au/travel/blogs/yowie-man/tim-the-yowie-man-the-Bunyip-hunt-20131122-2y0o8.html 22 November.
- Wagner, J. and Brooks, R. (1973). *The Bunyip of Berkeley's Creek.* Longman Young (Melbourne).
- Waldron, D. and Townsend, S. (2012). *Snarls From the Tea-Tree: Big Cat Folklore.* Arcadia (Victoria).
- Williamdefalco (2014). Cryptids and monsters: The bunyip. YouTube, https://www.youtube.com/watch?v=GtgWTCADfFM uploaded 12 February.
- Wrightson, P. and Horder, M. (1973). *The Bunyip Hole.* Hutchinson (Richmond).
- Young, Bob (1977) Music soundtrack for *Dot and the Kangaroo.* Sweet Soundtrack.

A CHECKLIST OF HISTORICAL HYPOTHESES FOR THE LOCH NESS MONSTER

C.G.M. Paxton[1] and A.J. Shine[2]
[1]Centre for Research into Ecological and Environmental Modelling,
University of St Andrews,
The Observatory, Buchanan Gardens,
St Andrews, Fife KY16 9LZ, Scotland,
United Kingdom
and
[2]Loch Ness Project,
Drumnadrochit, Inverness-shire IV63 6UW,
Scotland,
United Kingdom

ABSTRACT
This paper provides a comprehensive checklist of hypotheses concerning the putative identity of the Loch Ness Monster that have been offered to date. Hoaxes and specific identities for land sightings are not included. No effort is made to review or evaluate the hypotheses.

INTRODUCTION
Over 80 hypotheses for the Loch Ness Monster have been suggested (see below). Most of these hypotheses were suggested in the period 1930–1934 primarily in Scottish newspapers (see below). Subsequently, many of these hypotheses have been re-hypothesised (see tables below), presumably because earlier explanations were unknown to the originators. To clarify the attribution of hypotheses, a checklist is provided of the

earliest accounts of explanation that we have managed to find, along with prominent subsequent repeat suggestions or useful discussions. We make no effort to review or evaluate hypotheses here nor do we suggest a given hypothesis reflects an actual belief by the originator.

EXPLANATIONS

We do not consider hoaxes or specific explanations for terrestrial reports, although we note that many of the explanations given in the tables below have also been applied to terrestrial accounts. In addition, goats have been suggested for some terrestrial accounts (e.g. Raynor, n.d.). We draw a distinction between zoological simile and actual explanation. So, for example, the possibly apocryphal report of Duncan Macdonald of a frog-like animal in the lake (Anon., 1934e) is not considered as an actual hypothesised giant frog. However, the distinction can sometimes be unclear. For example, witness George Spicer (Anon., 1933e) said that his animal was the "nearest approach to a dragon or pre-historic animal" that he had ever seen. Some suggestions (e.g. Arthur Keith, see below) appear to be made tongue in cheek. Also, note that similar suggestions were made for aquatic monsters reported from earlier periods. During the early 1800s, for example, Sir Walter Scott suggested an otter as the origins of a monster reported from Cauldshiels Loch (Scott, 1933).

Magin (2011) provides a detailed history of the Loch Ness Monster in the early 1930s, but we have found some references that predate his first found usages. The earliest reference that we have found to the phenomena in Loch Ness being referred to as a "monster" in English comes from 3 September 1930 (a "monster fish or animal" - Invernessian, 1930). The first reference that we have found in modern usage, i.e. "Loch Ness "Monster", comes from the *Aberdeen Journal* on 23 May 1933 (Anon., 1933a). The first use of "Nessie" for the monster that we have found was in the *Edinburgh Evening News* on 9 January 1934, p. 2 (Anon., 1934b). It was noticed as early as 26 September 1933 (Anon., 1933h) that the monster predominantly occurred in "calm and bright" conditions. The first reference to the Loch Ness Monster appearing as an upturned boat (direct quote "a boat upside-down"), a description pretty much unique to Loch Ness, is from the *Hartford Courant* on 12 October 1930 (reference in Magin, 2011), although this American paper is presumably quoting from an unknown British source.

Table 1: Natural physical phenomena as explanations for the Loch Ness Monster

Explanation	*Who*	*Earliest Reference (plus notable subsequent references)*	*Comments*
Earthquake-induced disturbances/ gas bubbling	Anon. "S." L. Piccardi Roy P. Mackal	Anon. (1933b). S. (1950). Anon. (2001), see also Piccardi (2001). Mackal (1976, p. 201).	
Geyser on bottom of lake	"A."	A. (1933).	Humorous suggestion.
Mirages	Alex Campbell	Alex Campbell's letter to employers, 28 October 1933, quoted in Gould, (1934, p. 110, see also p. 108).	Lehn (1979).
Pressure waves caused by rain run off	Correspondent	Anon. (1947).	
Reflections	Blackpool resident	Gould (1934, p. 108).	
Reflections/wind slicks (i.e. isolated areas of calm waters reflecting dark hillside)	Adrian Shine	This paper (see also Shine, n.d).	
Shadows	E.G. Boulenger	Anon. (1938).	
Standing waves*	Peter Baker and Mark Westwood Steuart Campbell	Baker and Westwood (1962). Campbell (1991, pp. 19, 27, 111).	Could be caused by wind, currents or wakes.
Water spouts	"M.A.F."	M.A.F. (1937).	
Wind waves (cat's paws etc.)	"W.L.F" W. Cranston Maurice Burton	W.L.F. (1933). Gould (1934, p. 110). Burton, (1961, plates 1 and 2).	

* Burton (1961) mentions standing waves as a proposal made by "authoritative" others. We have been unable to find these "others".

Internal seiches may account for surface phenomena moving against the wind (Shine and Martin, 1988).

Table 2: Human artefacts as explanations for the Loch Ness Monster

Explanation	*Who*	*Earliest Reference (plus notable subsequent references)*	*Comments*
Boats	"W.L.F" Maurice Burton	W.L.F. (1933). Burton (1961, p.73).	Implicit.
Artificial debris	Rupert Gould	Gould (1934, p. 107).	Gould mentions tar barrels.
Dummy mines deployed in 1918	Experts of Royal Naval School	Anon. (1950).	
Preserved viking boat of variable buoyancy	Unknown George F. Campbell	Unknown newspapers. Campbell (2006, pp. 53-59).	Meredith (1977, p. 15).
Submarines (as explanation rather than description)	"W.L.F."	W.L.F. (1933).	
Upturned boat	Anon. Well known Glasgow citizen	Anon. (1933d) *Daily Record and Mail* 4 November 1934* p. unknown.	But the Loch Ness monster was *described* as looking like an upturned boat before this.
Wake effects	Correspondent* Rupert Gould Peter Baker and Mark Westwood	*Daily Telegraph* 14 December 1933, p. unknown. Gould (1934, pp. 108-109). Baker and Westwood (1962).	Wakes can also be produced by animals, see Burton (1961, p. 105).

* Cited in Gould (1934) but we have been unable to look at the primary reference.

Table 3: Botanical explanations for the Loch Ness Monster

Explanation	*Who*	*Earliest Reference (plus notable subsequent references)*	*Comments*
Logs/trees/natural debris	Photo	Anon. (1933g).	
	W.G. Ramsay-Fairfax	Ramsey-Fairfax (1933).	
	Anon.	Anon. (1933n).	
	E.G. Boulenger*	*Observer* 29 October 1933, p. unknown.	
	Photo	Anon. (1933q).	
	Col. W.H. Lane and G. Davis	Gould (1934, p. 107).	
	Maurice Burton	Burton (1961, pp. 100-101).	
	Spokesman of Woodland Trust	Anon. (2014).	
Peat/vegetation mats	Prof. D.M Watson	Anon. (1933n)	
	E.G. Boulenger*	*Observer* 29 October 1933, p. unknown.	
	Maurice Burton	Burton (1961, pp. 91-103).	

* Cited in Gould (1934) but we have been unable to look at the primary reference.

Table 4: Zoological explanations for the Loch Ness Monster

Explanation	*Who*	*Earliest Reference (plus notable subsequent references)*	*Comments*
A salmon	"W.L.F."	W.L.F. (1933).	
Shoal of salmon	"Clachnaharry"	Clachnaharry (1933).	
Sturgeon	F. Sutherland "Piscator" Rupert Gould Adrian Shine	Anon. (1933d). Piscator (1933). Gould (1933). Shine (1993).	
Crane	?*	*Hexham Courant* 9 December 1933, p. unknown.	
Waterfowl	Alex Campbell	Letter (7 September 1933, in Gould, 1934, pp. 110-111).	
Seagull splashing water on its back	"W.L.F."	W.L.F. (1933).	
Otters	"Another angler" Anon. R. Elmhirst* Mr Gillies	An Angler (1930). Anon. (1933m) *Daily Mail* 27 October 1933, p. unknown. Letter (4 March 1934, in Gould, 1934, pp. 74 and 116).	See also the *Glasgow Herald* 5 January 1934, p.7.
Seal	"Not An Angler" Anon. Capt. D.J. Munro Rupert Gould Anon. Scientists Members of the Linnean Society	Not An Angler (1930). Anon. (1933m). Munro (1933). Gould (1933). Anon. (1934h). Anon. (1934g). Anon. (1934i).	There is a note of seal in River Ness by "R.A.M." (*Northern Chronicle* 10 September 1930, p. 5); seals have been known in the River Ness since at least c. 1720s (Anon., in fact Burt, 1754, p. 43, Stevenson 2004).
Cow (dead)	"GLORAT."	GLORAT. (1934).	See also the *Dundee Courier* 30 December 1933, p. 5.
Deer	Maurice Burton	Burton (1961, pp. 131-135).	
Horse	"Equinox"	Equinox (1933).	

* Cited in Gould (1934) but we have been unable to look at the primary reference.

Table 5: Cryptozoological explanations (animals outside their known ranges in space and time) for the Loch Ness Monster

Explanation	*Who*	*Earliest Reference (plus notable subsequent references)*	*Comments*
a) unexpected in space			
Cuttlefish/squid (giant) (see also novel animals section)	"Delta" William Beebe David Stead	Delta (1933). Anon. (1934d). Anon. (1934f).	All known cephalopods are marine.
Generic shark	D. Murray Rose	Murray Rose (1933).	Mostly marine.
Greenland shark *Somniosus microcephalus*	Jeremy Wade	Weston (2013).	Not known in freshwaters but is known from the Scottish coast, see *Scotsman* 19 January 1929, p. 10.
Frilled shark *Chlamydoselachus anguineus*	Roy P. Mackal	Mackal (1976, pp 140-141).	Not known in freshwaters.
Basking shark *Cetorhinus maximus***	Anon. A fellow of the Zoological Society (Scotland) James Ritchie	Anon. (19331). A fellow of the Zoological Society (Scotland) (1933) Ritchie (1933).	Not known in freshwaters. Not wholly clear if he was actually advocating it as an explanation.
Ray/skate	"W.L.F." Paul and Lena Bottriell	W.L.F. (1933). Taylor (1988).	North Atlantic species. All marine.
Manta ray *Manta birostris*	"Delta" Robert Fleming***	Delta (1933). *Daily Record* 2 January	Manta rays are marine and not known in temperate

		1934, p. unknown.	waters.
Arapaima *Arapaima gigas*	"Delta"	Delta (1933).	
Conger eel *Conger conger*	"Not An Angler" Editorial Harold Frere	Not An Angler (1930b). Anon. (1933i). Frere (1933).	Not known in freshwaters.
Catfish (wels *Siluris glanis*?)	Dr A van Vieldhuizen***	*Daily Mail* 5 January 1934, p. unknown.	
Giant sunfish *Mola mola*	Allan McLean	McLean (1933).	Not known in freshwaters. The account confusingly mentions Australia, where sunfish can refer to the basking shark, but *Mola mola* is meant.
Sailfish *Istiophorus* sp.	Anon.	Anon. (1933l).	Not known in freshwaters. Maybe confused with basking shark, see Anon. (1933c).
Swordfish *Xiphias gladius*	"A."	A. (1933).	
Oarfish (ribbonfish) *Regalecus glesne*	H.R. Jukes	Jukes (1933).	Not known in freshwaters.
Crocodile	"A." Miss J.S. Fraser*** "A.B.S."	A. (1933). *Daily Mail* 1 January 1934, p. unknown. A.B.S. (1934).	Humorous suggestion? Not part of European fauna.
Tortoise/turtle	Dr Barnett*** Major	*Daily Mail* 1 January 1934, p. unknown. *Daily*	The only turtle species (leatherback *Dermochelys*

	Meikle***	*Record and Mail* 8 December 1934, p. unknown.	*coriacea*) that regularly comes into UK waters is not known to ever enter freshwaters. .
Sea snake	H.R. Jukes	Jukes (1933).	
Anaconda *Eunectes* sp.	Editorial	Anon. (1933i).	From South America.
Brazilian giant otter *Pteronura brasiliensis*	?***	*Daily Telegraph* 14 December 1933, p. unknown.	Presumed escapee.
Hippopotamus *Hippopotamus amphibius*	Sinclair Payne	Payne (1934).	Humorous?
Elephant	Walter Wood Dennis Power and Donald Johnson Neil Clark	Wood (1934). Power and Johnson (1979). Clark (2005).	Maybe humorous, does not directly mention Loch Ness.
Generic whale	Special correspondent L. Gordon Grant	Special Correspondent (1933). Gordon Grant (1934).	
Fin whale *Balaenoptera physalus*	Anon.	Anon. (1933p).	
Killer whale *Orcinus orca*	Dr A Fraser-Brunner***	*Daily Mail* 23 April 1934, p. unknown.	
Sowerby's beaked whale *Mesoplodon bidens*	Prof. Zuckmeyer	*Exchange* magazine.	Reprinted *Evening Dispatch* (Edinburgh) 15 January 1934, p. 4.
Narwhal *Monodon monoceros* (female), giant beaked whale *Berardius* sp., beaked whale *Mesoplodon* sp., Cuvier's beaked whale *Ziphius cavirostris*, and	Rupert Gould	Gould, (1934, pp. 133-134).	

humpback whale *Megaptera novaeangliae*			
Bottlenose whale *Hyperoodon* sp.	Anon.	Anon. (1933r).	Sources refer to bottlenose whales exclusively, but it may be that bottlenose dolphin *Tursiops truncatus* is being referred to as these are common in the Moray Firth.
Beluga (aka white) whale *Delphinapterus leucas*	Percy H. Grimshaw Rupert Gould	Grimshaw (1933). Gould (1933).	The Loch Ness monster is seldom described as "white". A white whale was found up the Firth of Forth, *Scotsman* 15 October 1932, p. 12.
Porpoise *Phocoena phocoena*	"Not An Angler" "Horribly Matter of Fact"	Not An Angler (1930). Horribly Matter of Fact (Fact, 1933).	
Generic dolphin	"16-bore" L. Gordon Grant	16-bore (1933). Gordon Grant (1934).	Specifically dolphins in a line.
"Grampus" (= Risso's dolphin *Grampus griseus*?, false killer whale *Pseudorca crassidens*, or killer whale)	"D.M."	D.M. (1933).	
Walrus *Odobenus rosmarus*	W.U. Goodbody	Anon. (1934a)**** Halcrow XXI	Tongue in cheek?

	"Halcrow XXI"	(1934).	
Camel	"Kelpie"	Kelpie (1933).	
Moose *Alces alces*	Dale Drinnon	Drinnon (2009).	
b) unexpected in time			
Marine saurian/prehistoric reptile/ *Plesiosaurus*	Anon. (Alex Campbell?) Mr Russell-Smith Special correspondent Rupert Gould	Anon. (1933f). Russell-Smith (1933). Special Correspondent (1933). Gould (1934, pp. 119-124).	Reprinted *Inverness Courier* 10 October 1933, p. 4.
Archaeocete whale	Roy P. Mackal	Wolfinger (1998).	
b) unexpected in space and time			
Sirenian (Steller's sea cow *Hydrodamalis gigas*)	Dennis Bickmore Roy P. Mackal	Bickmore (1934). Mackal (1976, p. 136).	Previously suggested for Lake Okanagan reports, e.g. *Vancouver Province*, 25 July 1926, p. 1.

*i.e. novel animals or animals whose range is not recognised as including Loch Ness in time or space.

**Gould has states a basking shark was suggested in the *Scotsman* on 20 October 1933. He is in error, the relevant letter was published on 21 October and does not actually state that the LNM was a basking shark.

***Cited in Gould (1934) but we have been unable to look at the primary reference.

****Goodbody is not advocating the hypothesis but refers to it as an already suggested hypothesis.

Table 6: Novel cryptozoological animals as explanations for the Loch Ness Monster

Explanation	*Who*	*Earliest Reference (plus notable subsequent references)*	*Comments*
Opisthobranchia	Roy P. Mackal	Mackal (1976, p. 143).	
Endemic(?) freshwater giant cuttlefish/squid	Tony Shiels	Shiels (1984).	All known cephalopods are marine.
A *Tullimonstrum*-like giant invertebrate	F.W. Holiday	Holiday (1968).	
Crab (giant)	"A."	A. (1933).	Humorous suggestion?
Eel (giant)*	Anon. Editor Norman Morrison John F. Burton Copenhagen correspondent "Kelpie"	Anon. (1933b). Anon. (1933h). Anon. (1933o). Burton (1934). Copenhagen Correspondent (1934). Kelpie (1934).	With reference to leptocephalus larvae.
Eel (giant) *Anguilla anguilla*	Roy P. Mackal Richard Freeman	Cited in Stark (1977). Barker (2003).	Specifically an *A. anguilla* that grows large in freshwaters rather than returns to the sea to spawn.
Giant salamander/amphibian	George Spicer W.H. Lane Roy P. Mackal	Anon. (1933j). Anon. (1933k). Mackal (1976, p. 139).	Not wholly clear if Spicer was advocating an actual amphibian or an amphibious animal. See also (Lane, 1934).
"Plesioturtle"	Kenneth Carpenter?	Starkey (2008).	An evolved long-necked leatherback turtle.
Short-necked plesiosaur	Kenneth Carpenter	Starkey (2008).	Not a pliosaur but an animal evolved from long-necked plesiosaurs.

Sea serpent (i.e. large marine snake)	D.J. Munro	Anon. (1934c).	
Seal ("huge")	"Camper"	Camper (1930).	
Long necked giant pinniped	Anton C. Oudemans	Oudemans (1934).	
Giant otter	?** Maurice Burton	*Daily Record and Mail* 27 January 1934, p. unknown. Burton (1961, p. 166).	

*It is sometimes unclear if writers are advocating a bona fide huge eel or simply a misinterpretation of a slightly larger than average eel.

** Cited in Gould (1934) but we have been unable to look at the primary reference.

Table 7: Other explanations for the Loch Ness monster

Explanation	*Who*	*Earliest Reference (plus notable subsequent references)*	*Comments*
Ghost/apparition	Sir Arthur Keith* F.W. Holiday	*Daily Mail* 3 January 1934, p. unknown. Holiday (1973).	Humorous? Not wholly clear.
Generated apparition (tulpa)	Roland Watson	Hepple (1982).	
Extraterrestrial	Warren Smith	Smith (1976).	
Hallucinations	"W.L.F."	W.L.F.(1933).	

* Cited in Gould (1934) but we have been unable to look at the primary reference.

ACKNOWLEDGEMENTS
Our thanks to Gordon Rutter, Karl Shuker and Roland Watson for bibliographic help and to referees for comments.

REFERENCES

- 16-bore (1933). Letter. *Field,* 162: 1371.
- "A". (1933). The monster – give it a name. *Dundee Courier* (Dundee), 2 November: 6.
- A Fellow of the Zoological Society (Scotland) (1933). Letter. *Inverness Courier* (Inverness), 24 October: 5.
- A.B.S. (1934). Letter. *Inverness Courier* (Inverness), 12 June: 3.
- An Angler (1930). Letter. *Northern Chronicle* (Inverness), 10 September: 5.
- Anon. [Burt, E.] (1754). *Letters from A Gentleman in the North of Scotland to His Friend in London.* J. Birt (London).
- Anon. (1933a). Loch Ness "Monster" again seen. *Aberdeen Journal* (Aberdeen), 23 May: 8.
- Anon. (1933b). Strange spectacle on Loch Ness. *Inverness Courier* (Inverness), 23 May: 4.
- Anon. (1933c). A monster shark. *Scotsman* (Edinburgh), 7 June: 9.
- Anon. (1933d). That Loch Ness "Monster"! *Dundee Courier* (Dundee), 16 June: 5.
- Anon. (1933e). Loch Ness Monster. *Inverness Courier* (Inverness), 4 August: 5.
- Anon. (1933f). Loch Ness Monster. *Northern Chronicle* (Inverness), 9 August: 5.
- Anon. (1933g). Untitled. *Aberdeen Journal* (Aberdeen), 16 August: 7.
- Anon. (1933h). Editorial. *Inverness Courier* (Inverness), 26 September: 4.
- Anon. (1933i). Editorial. *Northern Chronicle* (Inverness), 27 September: 4.
- Anon. (1933j). Loch Ness 'Monster'. *Inverness Courier* (Inverness), 3 October: 5.
- Anon. (1933k). "Species of salamander". *Inverness Courier* (Inverness), 10 October: 4.
- Anon. (1933l). Loch Ness "Monster". *Northern Chronicle* (Inverness), 11 October: 5.
- Anon. (1933m). Loch Ness Monster. *Scotsman* (Edinburgh), 17 October: 9.
- Anon. (1933n). Loch Ness Monster. *Scotsman* (Edinburgh), 18 October: 11.
- Anon. (1933o). Loch Ness Monster. *Glasgow Herald* (Glasgow), 6 December: 13.
- Anon. (1933p). Is the Monster a fin whale? *Aberdeen Journal* (Aberdeen), 7 December: 8.
- Anon. (1933q). Loch Ness Monster. *Bulletin and Scots Pictorial* (Glasgow), 12 December: 17.
- Anon. (1933r). The "Monster" mystery. *Lancashire Evening Post* (Preston), 13 December: 4.
- Anon. (1934a). Long, thin neck and small head. *Scotsman* (Edinburgh), 1 January: 15.
- Anon. (1934b). Loch Ness film. *Edinburgh Evening News* (Edinburgh), 9 January: 2.
- Anon. (1934c). Loch Ness monster. *Inverness Courier* (Inverness), 16 January: 4.
- Anon. (1934d). A giant squid. *Scotsman* (Edinburgh), 27 January: 15.
- Anon. (1934e). A diver's experience – a strange creature – strange creatures. *Northern Chronicle* (Inverness), 31 January: Not known.
- Anon. (1934f). Loch Ness Monster. *Scotsman* (Edinburgh), 14 September: 8.
- Anon. (1934g). Untitled. *Glasgow Herald* (Glasgow), 5 October: 10.
- Anon. (1934h). Editorial. *Nature,* 133: 56.

- Anon. (1934i). Untitled. *Nature,* 133: 56.
- Anon. (1938). Loch Ness Monster. *Scotsman* (Edinburgh), 22 August: 13.
- Anon. (1947). A Scotsman's log. *Scotsman* (Edinburgh), 23 July: 4.
- Anon. (1950). Highlands ablaze in defence of 'Nessie'. *Aberdeen Journal* (Aberdeen), November: 5.
- Anon. (2001). Quake theory shakes legend of Loch Ness. *Times* (London), 27 June: 5.
- Anon. (2014). Has the mystery of the 'Log Ness Monster' been solved? *Independent* (London), 22 November: Not known.
- Baker, P. and Westwood, M. (1962). Sounding out the monster. *Observer* (London), 26 August: 9.
- Barker, W. (2003). Nessie is old eunuch eel. *Sun* (London), http://www.thesun.co.uk/sol/homepage/news/144039/Nessie-is-old-Eunuch-Eel.html 22 September: Not known.
- Bickmore, D. (1934). Letter. *Aberdeen Journal* (Aberdeen), 3 January: 2.
- Burton, J.F. (1934). Letter. *Times* (London), 5 January: 8.
- Burton, M. (1961). *The Elusive Monster.* Rupert Hart-Davis (London).
- Campbell, G.F. (2006). *The First and Lost Iona.* Candlemas Hill (Glasgow).
- Campbell, S. (1991). *The Loch Ness Monster: The Evidence.* Aberdeen University Press (Aberdeen).
- Camper (1930). Letter. *Inverness Courier* (Inverness), 3 September: 5.
- Clachnaharry (1933). Letter. *Evening Dispatch* (Edinburgh), 27 November: 6.
- Clark, N.L.G. (2005). Tracking dinosaurs in Scotland. *Journal of the Open University Geological Society,* 26 (2): 30-35.
- Copenhagen Correspondent (1934). 6ft eel larvae. *Times* (London), 17 January: 12.
- D.M. (1933). Letter. *Scotsman* (Edinburgh), 23 September: 10.
- Delta (1933). Letter. *Northern Chronicle* (Inverness), 1 November: 3.
- Drinnon, D. (2009). Origins of the water horse. Still On The Track, http://forteanzoology.blogspot.co.uk/2009/06/dale-drinnon-origins-of-water-horse.html 22 July.
- Equinox (1933). Letter. *Dundee Courier* (Dundee), 24 October: 3.
- Frere, H. (1933). Letter. *Inverness Courier* (Inverness), 3 October: 5.
- GLORAT. (1934). The Loch Ness Monster. *Falkirk Herald* (Falkirk), 17 March: 5.
- Gordon Grant, L. (1934). Letter. *Scotsman* (Edinburgh), 1 January: 11.
- Gould, R.T. (1933). The Loch Ness "Monster". *Times* (London), 9 December: 13.
- Gould, R.T. (1934). *The Loch Ness Monster.* Geoffrey Bles (London).
- Grimshaw, P.H. (1933). Letter. *Scotsman* (Edinburgh), 28 October: 12.
- Halcrow XXI (1934). Letter. *Scotsman* (Edinburgh), 20 January: 18.
- Hepple, R. (1982). Untitled. *Nessletter,* No. 55: 1.
- Holiday, F.W. (1968). *The Great Orm of Loch Ness.* Faber and Faber (London).
- Holiday, F.W. (1973). *The Dragon and The Disc.* Sedgewick and Jackson (London).
- Horribly Matter of Fact (1933). Letter. *Scotsman* (Edinburgh), 29 November: 13.
- Invernessian (1930). Letter. *Northern Chronicle*, 30 September: 5.
- Jukes, H.R. (1933). "Monsters" seen in past. *Aberdeen Journal* (Aberdeen), 26 October: 6.

- Kelpie (1933). Letter. *Scotsman* (Edinburgh), 4 November: 15
- Kelpie (1934). Letter. *Field,* 163: 78.
- Lane, W.H. (1934). *The Home of the Loch Ness Monster.* Moray Press (Edinburgh).
- Lehn, W.H. (1979). Atmospheric refraction and lake monsters. *Science,* 205: 183-185.
- M.A.F. (1937). Letter. *Inverness Courier* (Inverness), 24 September: 5.
- Mackal, R.P. (1976). *The Monsters of Loch Ness.* Swallow Press (Chicago).
- Magin, U. (2011). *Investigating the Impossible: Sea-Serpents in the Air, Volcanoes that Aren't, and Other Out-of-Place Mysteries.* Anomalist Books (San Antonio).
- McLean, A. (1933). Letter. *Scotsman* (Edinburgh), 16 November: 11.
- Meredith, D.L. (1977). *Search at Loch Ness.* Quadrangle/The New York Times Book Company (New York).
- Munro, D.J. (1933). Letter. *Scotsman* (Edinburgh), 1 November: 13.
- Murray Rose, D. (1933). Letter. *Scotsman* (Edinburgh), 20 October: 11.
- Not An Angler (1930). Letter. *Northern Chronicle* (Inverness), 3 September: 5.
- Oudemans, A.C. (1934). *The Loch Ness Animal.* E.J. Brill (Leyden).
- Payne, S. (1934). Letter. *Field,* 163: 29.
- Piccardi, L. (2001). Seismotectonic origins of the monster of Loch Ness. Paper presented at the Earth System Processes - Global Meeting (24-28 June 2001).
- Piscator (1933). Letter. *Northern Chronicle* (Inverness), 16 August: 5.
- Power, D. and Johnson, D. (1979). A fresh look at Nessie. *New Scientist,* 83: 358-359.
- Ramsey-Fairfax, W.G. (1933). Letter. *Morning Post* (London), 10 October: 6.
- Raynor, D. (n.d). Loch Ness Investigation, http://www.lochnessinvestigation.com/Goats.html Retrieved 15 November 2015.
- Ritchie, J. (1933). Letter. *Times* (London), 16 December: 15.
- Russell-Smith (1933). Letter. *Morning Post* (London), 6 October: 15.
- S. (1950). Letter. *Scotsman* (Edinburgh), 16 September: 6.
- Scott, W. (1933) *The Letters of Sir Walter Scott Vol. 4.* Constable (London).
- Shiels, A. (1984). Mother nature's jumbo jet. *Fortean Times,* No. 42: 62–67.
- Shine, A.J. (1993). Postscript: surgeon or sturgeon? *Scottish Naturalist,* 105: 271-282.
- Shine, A.J. (n.d). Loch Ness and Morar Project, http://www.lochnessproject.org/loch_ness_reflections_news_links/sightingkey/LOCH_NESS_SIGHTING_3.HTM Retrieved 4 June 2015.
- Shine, A.J. and Martin, D.S. (1988). Loch Ness habitats observed by sonar and underwater television. *Scottish Naturalist,* 105: 170.
- Smith, W. (1976). *UFO Trek.* Zebra (New York).
- Special Correspondent (1933). Loch Ness Monster. *Scotsman* (Edinburgh), 16 October: 11.
- Stark, W.G. (1977). *In Search of the Loch Ness Monster.* Alan Landsburg Productions (U.S.A.).
- Starkey, D. (2008). The Loch Ness Monster revealed. Discovery Communications (Bethseda).

- Stevenson, D. (2004). Burt, Edmund (d. 1755). Oxford Dictionary of National Biography. http://www.oxforddnb.com/view/article/4118 accessed 23 May 2015, retrieved 26 November 2015.
- Taylor, D. (1988). Loch Ness mystery solved! *High Wycombe Star* (High Wycombe), 28 October: 1-2.
- W.L.F. (1933). Letter. *Aberdeen Journal* (Aberdeen), 18 December: 2.
- Weston, D. (2013). *River Monsters: The Legend of Loch Ness.* Icon Films (London).
- Wolfinger, K. (1998). *The Beast of Loch Ness.* Nova WGBH (Boston).
- Wood, W. (1934). Letter. *Times* (London), 8 January: 8.

NOT FINDING BIGFOOT IN DNA

Haskell V. Hart
Canyon Lake, TX 78133, USA
Email: hvhart@swbell.net

ABSTRACT
Over the past 200 years, there have been tens of thousands of alleged sightings of tall, hairy, bipedal primates all over North America, and they are found in oral legends and art of many Native Americans tribes. However, there is presently no holotype specimen or even any verified skeletal remains or fossils that would substantiate these claims. Ketchum *et al.* (2013) published purported bigfoot/sasquatch DNA sequences from hair, flesh, saliva, and blood samples of unknown provenance (no animal was seen to have deposited the samples). They concluded that the samples came from a hybrid between an unknown primate male and a modern *Homo sapiens* female. Three of these nuclear DNA sequences are reexamined here and found to match database and literature sequences of known species of mammals: a black bear, a human, and a dog.

INTRODUCTION
Large, hairy, human-like figures have been reported in nearly all areas of the world and known as sasquatch or bigfoot in North America, almasty, yeren, abominable snowman, and yeti in Asia, yowie in Australia, and hairy man or wild man nearly everywhere else (Sykes, 2015). However, the authenticity of these sightings has been widely debated, and there is no general acceptance of these claims by the scientific community.

Although previously purported to exist by both Europeans and Native Americans (Strain, 2008), the phenomenon of bigfoot received a boost in interest when Gerald Crew found and cast large human-like footprints near logging operations in northern California in 1958 (Sanderson, 1959). By far the most convincing video evidence was obtained in 1967, also in California, in the famous Patterson-Gimlin film. However, the authenticity of the latter is still debated, and the film has been meticulously analysed by, among others, experienced Hollywood costume designer and makeup artist Bill Munns (2014), who concluded that it could not be faked with the costume materials and technology available in 1967. Nevertheless, the existence of bigfoot continues to be hotly debated, mostly with regard to poor-resolution photographs and videos, audio recordings, and interminable personal accounts of "sightings" in nearly all 50 US states. Numerous organisations exist for the sole purposes of documenting and following up on these alleged sightings (now in the tens of thousands) and searching for the elusive creature known as bigfoot or sasquatch. However, at present, no holotype specimen has been obtained, and no skeletal remains or fossils have ever been discovered and interpreted by science to be bigfoot.

Then in February, 2013, the much-anticipated sasquatch DNA paper appeared as the only paper in the first (and so far the only) issue of a journal owned by the principal author, Dr Melba Ketchum, DVM (Ketchum *et al.*, 2013). The paper had nine other co-authors, from a variety of fields. Finally, it was hoped, the existence of bigfoot might be proven. But, as will be seen, the search methodology and sequence interpretation make all the difference in cases of unknown species identification.

A sequence of the four nucleotide bases, represented by the letters A, T, G, and C, does not in itself prove anything (these bases are adenine, thymine, guanine and cytosine, which when paired A-T and G-C form the hydrogen bonds that hold together the two strands of DNA's double helix structure - Watson and Crick, 1953). It must be compared to reference sequences (i.e. more than one reference sequence) of known identities before any conclusions can be drawn from it.

This paper concerns those comparisons and the underlying methodology that guarantees that nothing is missed, at least as far as known and sequenced species are concerned. It is the first published reanalysis of the three Ketchum *et al.* (2013) nuclear DNA sequences.

COMPUTER METHODS

The nuclear DNA sequences of Samples 26, 31, and 140 were downloaded in FASTA format from the Ketchum *et al.* (2013) paper on the *DeNovo* journal website and were subsequently made available on the Sasquatch Genome Project website, http://www.sasquatchgenomeproject.org, as Supplemental Data 4, 5, and 6, respectively. Databases (collectively called GenBank) were queried with these entire sequences on the National Center for Biotechnology Information (NCBI) website (http://www.ncbi.nlm.nih.gov) with the use of their BLAST™- Basic Local Alignment Search Tool – software, (Altschul *et al.*, 1990; Madden, 2003) as was done by Ketchum *et al.* (2013). Default search parameters were used except that the word size was changed to 64 in order to accommodate the long sequences, and the number of reported hits was adjusted to fit the circumstances.

Additional match criteria were sometimes applied through the "Organism" input field or the Boolean logic language in the "Entrez Query" input field. For example, all Ursidae (bears) only. Otherwise, no additional or unusual search techniques were employed. The broader phylogeny was encompassed as well as limited searches to the most specific family, genus, or species possible. All relevant databases were searched in GenBank, not just the most popular "Nucleotide" database as Ketchum *et al.* (2013) did. In this way, absolutely no potential match was missed, and the most possible species were queried. Tables of hits were downloaded from the results page as Microsoft Excel® files for sorting and comparisons. Phylotrees (evolutionary tree of life) were constructed with the BLAST™ "Distance tree of results" option on the results page.

In the single case of the literature black bear data of Cahill *et al.* (2013), a PC standalone database of 2.1 Gb of black bear sequences (scaffolds) was constructed, and PC BLAST™ search software was downloaded from NCBI, due to the large number of data.

RESULTS AND DISCUSSION

Sample 26
Common NCBI Databases

1. Nucleotide (N), Transcriptome Shotgun Assembly (TSA), and Reference Genomic Sequences (RGS) Databases.

Sample 26 (S26) is a small piece of fur and attached tissue from California, found near a purported bigfoot kill, known as the Justin Smeja Kill, five weeks later. The main body was never found. The sequence contains 2,726,786 base pairs (bp).

The entire published S26 sequence was queried against GenBank in five separate species searches. Table 1 shows typical search results for one of the top 15 human hits by score, one of the top 15 Ursidae (bear family) hits, and hits over the same sequence range for other primates (OP), the genus *Canis* (dogs, wolves, coyotes), and all other (AO) species not previously included (each in a separate row). There were no black bear *Ursus americanus* matches over any of these 17 sequence ranges (13 coincided between human and bear), because there are much fewer black bear sequences than polar bear *U. maritimus* in these databases. In addition, the brown bear *U. arctos*, which includes grizzly bears as subspecies, does not have enough sequence data in GenBankfor valid comparisons; it was last seen in California in 1924. Ursidae were the best match to the S26 sequence over all 17 ranges (14 best matches were polar bear; two were giant panda *Ailuropoda melanoleuca*, one was tied). Human and other primates were the poorest matches throughout.

Please note that in this particular context, the term 'score' is a function of sequence length, mismatched bases, and gaps in alignment. See Madden (2003) for details. It should not be used as the sole match criterion, because a long sequence with lower %ID may outweigh a shorter sequence with very good or even perfect (100) %ID.

Table 1:

Sample 26 Sequence Match Example (1 of 17 best hits)				
HitSequence Range 189026-191141				
S26 *vs.*	ACCESSION *(a)*	% ID	SCORE	Av. %ID *(b)*
human-*N* ***(c)***	BC038508.1	93.3	3136	94.7
OP ***(d)*** -*N*: chimpanzee	XM_003951836.1	93.2	3120	94.9
Canis: dog-*N*	NM_001097982.2	95.4	3297	96.0
Ursidae: polar bear-*RGS* ***(e)***	NW_007929448.1	98.8	3788	99.2
AO ***(f)*** -*N*: Pacific walrus	XM_004394587.1	96.6	3515	95.2
(a) In NCBI databases. ***(b)*** Over all 17 best hits. ***(c)*** Nucleotide database. ***(d)*** Other Primates. ***(e)*** Reference Genomic Sequences database. ***(f)*** All Other species.				

A phylotree of results based on an S26 query of the Reference Genomic Sequences Database is shown in Fig. 1. Clearly S26 is a bear, close to the polar bear and only slightly more distant from the giant panda, and in exactly the same position that a black bear would occupy according to the NCBI Taxonomy database (http://www.ncbi.nlm.nih.gov/taxonomy/).

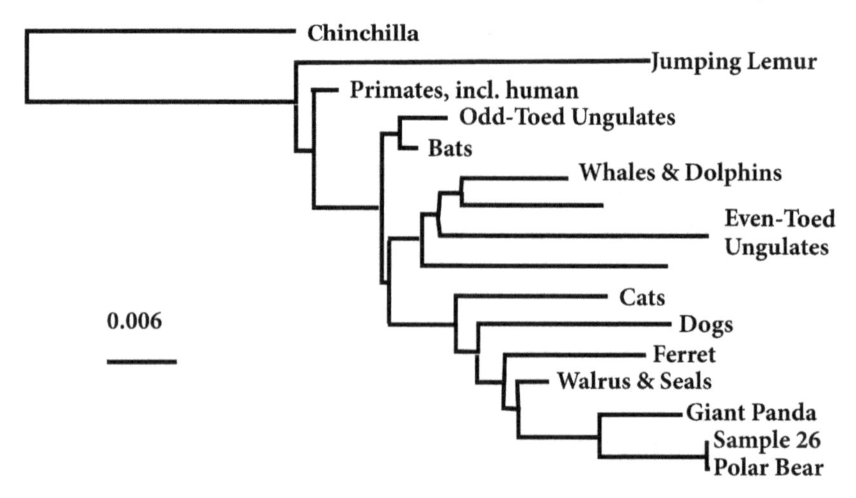

Fig. 1: Sample 26 and Mammals Phylotree

2. Expressed Sequence Tags (EST) Database

An expressed sequence tag is a nuclear cDNA (complementary DNA) sequence that is produced by cloning the corresponding mRNA sequence obtained from tissue extracts, usually from important organs such as eyes, heart, lung, kidney, liver, or muscle (Parkinson and Blaxter, 2009). Each sequence is typically a few hundred bp, which makes amplification and sequencing relatively easy compared to whole genome methods. These sequences are often highly conserved (as important organs function similarly among different species), so picking the "best" hit from among many "good" hits is critical.

Fortunately, the Expressed Sequence Tags database contains a large number (38,757) of mostly nonredundant black bear sequences. There are no other bear sequences in this database. By comparison, the more popular Nucleotide database, the only one considered by Ketchum *et al.* (2013), contains only 1663 black bear sequences, 216,511 giant panda, and 100,160 polar bear sequences as of this date (October 2015), reflecting the greater interest in the latter two as endangered species. Thus, it seemed like an unusual opportunity to search the EST database for black bear (and, of course, other) matches.

Overall results are summarised:

1. Of the best 29 black bear hits by score and %ID, 13 were **best** matched by black bear, 11 were best matched by polar bear sequences from the Reference Genomic Sequence database, and five were tied between these two bears. No other species were close.

2. Of the 30 best non-black bear hits, 27 were better matched over the same sequence range by polar bear. Two matches were equal between pig and polar bear and between many species and polar bear respectively. Only one best match of 59 total was not a bear but a dog. None of these 30 hit sequence ranges had any black bear data in the EST database. Sample 26 was found by a dog and may have been so contaminated.

Very clearly, S26 is again proven to be a bear, most likely a black bear.

3. RNA Reference Sequences (RNA ref_seq) Database

RNA (ribonucleic acid) is the complement of DNA (deoxyribonucleic acid): bases U (uracil) for A or A for T and C for G or G for C.

RNA uses three-base codons for amino acids to manufacture proteins, the workhorses of the cell. Any biochemistry text will explain this in more detail. The take away point here is that DNA queries can be searched against RNA sequences – the software makes the base conversions mentioned above. Here the RNA ref_seq database is queried with the entire S26 nDNA sequence as before. This database contains polar bear sequences but not black bear. A phylotree was produced from the BLAST™ results as Fig. 2 (carnivore section only). This is virtually identical to the opened version of Fig. 1. It clearly shows S26 in a close phylogenetic relationship to other carnivores, especially bears, and in the correct place for a black bear. The complete mammalian phylotree showed the very same animals as in Fig 1, including the phylogenetically distant primates, and with the same relationships as predicted by the NCBI Taxonomy database.

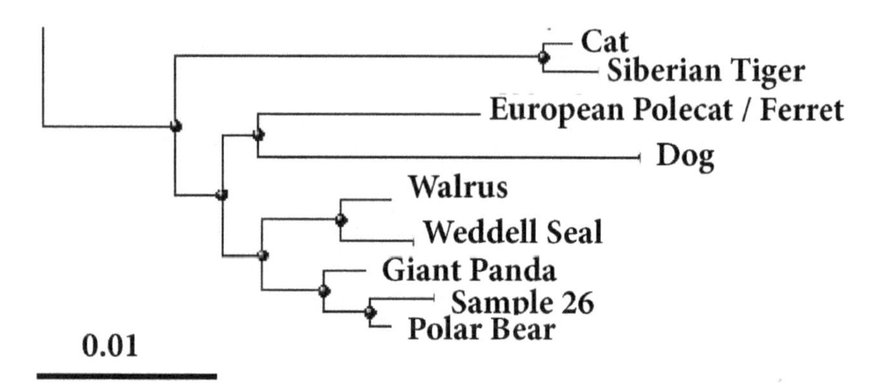

Fig. 2. Sample 26 and Carnivores Phylotree

Given the high interest in S26, the intense controversy over its provenance, and the paucity in these databases of sequences from black bear (the only extant bear species in California), it was decided to examine other sources for black bear nDNA sequences.

Comparisons to Literature Black Bear Data

1. Black Bear Genomic Scaffolds

The S26 nDNA sequence was queried with PC BLAST™ against the PC database of 238 black bear scaffolds from Cahill *et al.* (2013). Corresponding sequences of other species were retrieved from the NCBI Genomic Reference Sequences database for comparison. Table 2 shows a typical result of one of our sequence comparisons. Hits, top to bottom, each in single row, are numbered as:

1. S26 match to black bear data of Cahill *et al.*

The following hits are from queries of the Reference Genomic Sequences database.

2. Best S26 match to polar bear over same sequence range.

3. Best S26 match to giant panda over same sequence range.

4. Best S26 match to human over same sequence range.

5. Best S26 match to other primates (ex-human) over same sequence range.

When the top 50 hits (by score) were compared, 30 matched black bear best, eight matched polar bear best, 11 were tied between these two bears, and one was tied between polar bear and giant panda. Overall, black bear was the best species match to S26. All the human and other primate (OP) hits were significantly poorer matches than any bear matches. For conserved genes with the highest scores, we typically find that just a few percentage points % ID can be the difference between different higher taxa (e.g. Ursidae vs. Primates).

Very clearly, the black bear is the best match to S26.

Table 2:

Sample 26 Sequence Match Example (1 of 50 best hits)			
Hit Sequence Range 326857 - 328153			
S26 *vs.*	**ACCESSION**	**%ID**	**Av. %ID *(a)***
1. black bear	scaffold180 ***(b)***	99.8	99.9
2. polar bear	NW_007907318.1	99.7	99.8
3. giant panda	NW_003217713.1	99.0	99.2
4. human	NG_012881.1	95.2	95.8
5. OP: gorilla	NC_018435.1	95.0	95.8
(a) Over top 50 best bear hits and top 15 best other primate (OP) hits. ***(b)*** Scaffold number from Cahill *et al.* (2013) sequences.			

2. Ursidae UltraConserved Elements (UCE)

"UCE are highly conserved short DNA sequences [<700 bp here] that are shared by different organisms and are particularly useful for phylogeny estimation from genome sequence data." (Cronin *et al.*, 2014).

A UCE sequence text file was downloaded from the link in Cronin *et al.* (2014) and contained giant panda (3,985,224 bp), black bear (996,347 bp), brown bear (996,346 bp), brown bear – ABC Islands (996,346 bp), and polar bear (996,348 bp) sequences in FASTA format. The sequences were compared to S26 in its entirety with BLAST™, and a phylotree was constructed as Fig. 3. Clearly, it shows S26 in precisely the correct position for a black bear.

Fig. 3 is topographically identical to Fig. 3 in Cronin *et al.* (2014). The genetic group, brown bear – ABC, includes brown bears from the Admiralty, Baranof, and Chichagof (ABC) Islands off southeast Alaska, where significant past hybridisation of brown and polar bears has occurred (Cahill *et al.*, 2013; Cronin *et al.*, 2014).

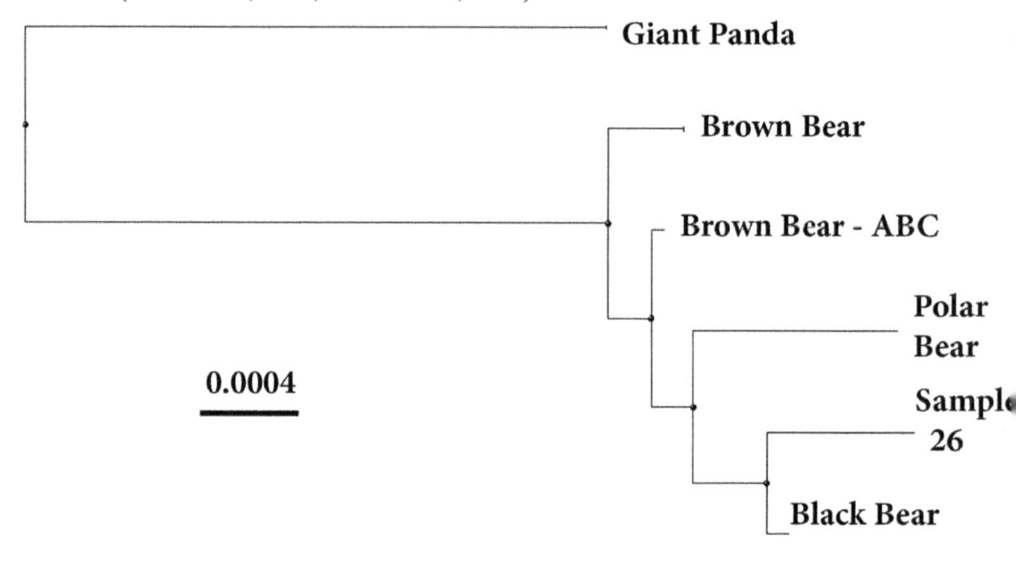

Fig. 3. Sample 26 and Bears Phylotree

Sample 31

Sample 31 (S31), from Alabama, came from a paper plate in a food trap intended to capture DNA of the animal(s) attempting to retrieve the food. No further details were given on the nature of the food or the trap. The sequence contains 527,761 bp.

As above for S26, the entire published S31 sequence was queried against GenBank in four separate searches: against human, all other primates, genus *Canis*, and all other species. The best hit was a fungus, indicating partial degradation of the sample. Of the 17 best hits, 12 were human and three were tied between human and chimpanzee, one was the fungus and one was probably an unknown bacterium. The best overall match is a human with 99.8%ID average, followed by other primates with 98.7%ID average. No other species even had consistent matches of any kind. Sample 31 is human.

The Ketchum *et al.* (2013) argument for hybridisation is based on their findings that the mitochondrial DNA, which is inherited through the matrilineal line (mother to daughter only), is basically human for 29 samples (but which critics say could represent contamination). Consequently, if the bigfoot or sasquatch is indeed an unknown primate-human hybrid as proposed by Ketchum *et al.* (2013), one would expect much more primate "best matches", especially considering the likely subsequent "breeding out" of the human component by pure unknown primate genes.

The results obtained above could only be achieved if F1 hybrid males and their male descendants continually mated with additional modern human females, in preference to the presumably more readily available unknown primate females. Conversely, breeding of unknown primate males with F1 hybrid females and their female descendants would eventually "breed out" the human component, contrary to what we observed. The alternate explanation, which seems unlikely, would entail a large influx of modern human male genes through continuous mating with F1 females, also requiring continuous modern human mothers. In any case, none of these scenarios can be proven from the S31 sequence, which is perfectly human and relatively short, compared to the human genome.

Sample 140

Sample 140 (S140) is blood from the inside of a gutter downspout in Illinois that had holes in the pattern of teeth marks and which was torn from a house, requiring considerable force. The sequence contains 2,101,957 bp.

As above for Sample 26, the entire published S140 sequence was queried against GenBank in four separate searches: against human, genus *Canis*, OP (other primates, i.e. non-human), and AO (all other species).

Table 3 shows one example of the top 15 *Canis* (dog, wolf, coyote) hits and the top 15 human hits in the Nucleotide database by score. Over all 17 of these sequence ranges (13 coincided) *Canis lupus familiaris* (the domestic dog) was the best match, better than human, other primates (OP), and all other (AO) species in the database. A phylotree, similar to Fig. 1 , was constructed, and S140 was within the *Canis* grouping, far removed from primates.

Table 3:

Sample 140 Sequence Match Example (1 of 17 best hits)				
Hit Sequence Range 655259-656836				
S140vs.	**ACCESSION**	**%ID**	**SCORE**	**Av. %ID *(a)***
Canis: dog	XM_540535.3	98.61	2789	98.5
human	NR_047674.1	94.99	2475	94.3
OP ***(b)***: chimpanzee	XM_001155521.3	95.12	2486	94.4
AO ***(c)***: Pacific walrus	XM_004404764.1	96.58	2612	95.3
(a) Over top 17 best hits. ***(b)*** OP = other primates. ***(c)*** AO = all other species.				

A final effort was made to find better mammalian matches for S140 by separately limiting hits to the order Carnivora, the families Canidae (dogs, wolves, coyotes, foxes), Felidae (cats), Mustelidae (weasels), Ursidae (bears), and Mephitidae (skunks), the clade Glires (rabbits and rodents), the tribe Sciurini (squirrels), the genus *Vulpes* (foxes), and these species - the house mouse *Mus musculus*, the common raccoon *Procyon lotor*, and the grey short-tailed opossum *Monodelphis domestica.* No matches were as good as *Canis lupus familiaris.*

CONCLUSIONS

The results presented in this paper show that nothing new was discovered by Ketchum *et al.* (2013); their published nDNA sequences of Samples 26, 31, and 140 are from known species.

The relevant conclusion from Ketchum *et al.* (2013) is:

"...the species [sasquatch] possesses a novel mosaic pattern of nuclear DNA comprising novel sequences that are related to primates interspersed with sequences that are closely homologous to humans."

Samples 26 and 140 contained no such sequences, being from a black bear and a dog respectively. Sample 31 is human, with no evidence of other primate sequences among the top hits.

The likely reasons for the above human-primate conclusion by Ketchum *et al.* (2013) are:

1. Only searching the Nucleotide database, which has limited black bear data and is dominated by human data.

The result is high scores with mediocre individual sequence %IDs for the many human hits, pushing the better %ID matches down in the hit list sorted by maximum score. No evidence of downloading the Excel™ hit list and sorting by %ID was given by Ketchum *et al.* (2013). At the time of the latter study, the only polar bear data was in the Transcriptome Shotgun Analysis (TSA) database, and was therefore overlooked.

2. Inexplicably, limiting searches to "human only" (Ketchum, 2014).

Even if a match is 100%ID, it is impossible to identify the species without comparisons. See 3. below.

3. Loose use of "homologous", without sufficient comparisons to other species.

Important genes, the ones most studied and sequenced, are usually the ones with the highest hit scores and are highly conserved in mammals. During this present study, it was even found that for some sequences polar bear and giant panda matched S26 equally (even 100%ID). Ketchum *et al.* (2013) provided no detailed search results or statistics to support their conclusions. Further, their phylotrees from nDNA search results made no sense in the context of hominin phylogeny. The only match comparisons of species in Ketchum *et al.* (2013) are in their phylotrees. Although Supplementary Figure 4 in Ketchum *et al.* (2013) is a rather normal-looking primate phylotree, Supp. Fig. 5 showed equally distant relationships to human and mouse and slightly more distant relationships to a chicken, a carp, 12 species of sharks, and 25 additional fish – but nothing else. Where are the primates and other mammals? Supp. Fig. 6 shows a relatively distant relationship to human and more distant to mouse – but nothing else. There was no indication of which figure corresponded with which sample. These two phylotrees are truncated or incomplete at best.

4. Using a human reference sequence to assemble the reads.

Because Ketchum *et al.* (2013) used a human Chromosome 11 reference sequence to assemble their reads into the three published supercontigs, the S26 (bear) and S140 (dog) supercontig sequences were *forced* to be similar to human (94-95%ID we found). Less conserved reads did not assemble, because they did not match the reference well enough. Thus, the published sequences are throughout highly conserved between bear (for S26) and human and between dog (for S140) and human. This makes discrimination between species more difficult. A better approach, considering the species to be unknown, would have been to do a *de novo* assembly, which does not use a reference sequence. Considering the entire S26 genome, the match to human would then have been only around 75%, clearly not good enough for identification as human.

Although the title of the Ketchum *et al.* (2013) paper claims sequencing of "three whole genomes", only about 0.1% of S26 and S140 genomes were sequenced, even less for S31. The whole human Chromosome 11 contains about 135 M (million) bp, again far more than the three published sequences (2.7 M, 0.5 M, and 2.1 M bp respectively). This is the result of using a human Chr. 11 reference sequence, which only assembled conserved genes.

Reference sequences are only correctly employed when the species is known (or very closely related) and one seeks only to identify a relatively small number (compared to the sequence length) of SNPs (single nucleotide polymorphisms – i.e. base pair differences). For unknown species, the correct approach is to do a *de novo* assembly, which makes no assumptions about the identity of the sample. This was a major mistake in methodology by Ketchum *et al.* (2013).

Positive results of the Ketchum *et al.* (2013) study include increased awareness of the potential of DNA analyses in cryptid identification. Their paper contained the first such attempt for a purported sasquatch/bigfoot, and the second attempt for any purported extant hominin. The first attempt was for a yeti, from which a hair sample proved to be from a horse (Malinkovitch *et al.*, 2004). A more recent comprehensive study utilised the mtDNA gene 12S rRNA to identify 30 hair samples, many quite well known and all thought to be from bigfoot or yeti (Sykes, 2014). All 30 were common mammals, including the Ketchum *et al.* (2013) Sample 26, which turned out to be a black bear. Additionally, two privately-funded DNA laboratory analyses of S26 also concluded that it was a black bear, contaminated by its collector. (Khan and White, 2012; Cassidy, 2013).

Also, as a result of the Ketchum *et al.* (2013) study and subsequent television shows (e.g. *The 10 Million Dollar Bigfoot Bounty* http://www.spike.com/shows/bigfoot-bounty), field researchers have learned the proper sample collection techniques required to minimise or exclude contamination.

Our conclusions, disappointing though they may be to some, do not disprove the existence of bigfoot. They simply mean that it is not to be found in these three samples; "absence of evidence is not evidence of absence". It is still possible that some day a documented (via photograph, video) DNA sample may be collected that shows a human-like DNA sequence distinct from all known primates. Even better would be whole body parts (a holotype) with their abundance of DNA. The search for evidence continues.

ACKNOWLEDGEMENTS

The author is grateful to James Cahill and Prof. Beth Shapiro of UC-Santa Cruz (California) for sharing their black bear data from Cahill *et al.* (2013). Also, thanks go to the Sasquatch Genome Project for making their DNA sequences available online.

No financial support was required or received for this work. Search software and database access are free online from NCBI. Online advice was promptly and appropriately provided by their helpdesk.

REFERENCES

- Altschul, S.F., *et al.* (1990). Basic local alignment search tool. *Journal of Molecular Biology*, 215 (No. 3): 403-410.
- Cahill, J.A., *et a*l. (2013). Genomic evidence for island population conversion resolves conflicting theories of polar bear evolution. *PLOS Genetics*, http://journals.plos.org/plosgenetics/article?id=10.1371/journal.pgen.1003345 14 March.
- Cassidy, B.G. (2013). *Technical Examination Report DNAS Case Number: 2012-006524*. DNA Solutions, Inc. (Oklahoma City).
- Cronin, M.A., *et al.* (2014). Molecular phylogeny and SNP variation of polar bears (*Ursus maritimus*), brown bears (*U. arctos*), and black bears (*U. americanus*) derived from genome sequences. *Journal of Heredity,* 105 (No. 3): 312–323.
- Khan, T. and White, B. (2012). *Final Report on the Analysis of Samples Submitted by Tyler Huggins: Case File 12*-019. Wildlife Forensic DNA Laboratory Trent University Oshawa (Peterborough, Canada).
- Ketchum, M.S., *et al.* (2013). Novel North American hominins, next generation sequencing of three whole genomes and associated studies. *DeNovo,* 1 (No. 1), http://sasquatchgenomeproject.org/linked/novel-north-american-hominins-final-pdf-download.pdf and also http://www.advancedsciencefoundation.org/#!novel-north-american-hominins/cayh
- Ketchum, M.S. (2014). Personal communication (e-mail), 16 January.

- Madden, T. (2003). The BLAST sequence analysis tool. *In:* McEntyre, J. and Ostell, J. (eds). *The NCBI Handbook.* National Center for Biotechnology Information (Bethesda). http://www.ncbi.nlm.nih.gov/books/NBK21097/
- Malinkovitch, M.C., *et al.* (2004). Molecular phylogenetic analyses indicate extensive morphological convergence between the "yeti" and primates. *Molecular Phylogenetics and Evolution*, 31: 1–3.
- Munns, W. (2014). *When Roger Met Patty: 1 Minute of Film...47 Years of Controversy.* Self-published and Kindle.
- Parkinson, J. and Blaxter M. (2009). Expressed sequence tags: an overview. *Methods in Molecular Biology*, 533: 1-12.
- Sanderson, I.T. (1959). The strange story of America's abominable snowman: "The Jerry Crew story". *True Magazine,* http://www.bigfootencounters.com/articles/true1959.htm December.
- Strain, K. M. (2008). *Giants, Cannibals and Monsters: Bigfoot in Native Culture.* Hancock House (Blaine).
- Sykes, B. (2014). Genetic analysis of hair samples attributed to yeti, bigfoot and other anomalous primates. *Proceedings of the Royal Society B*, 281: 161-163.
- Sykes, B. (2015). *The Nature of the Beast: The First Scientific Evidence on the Survival of Apemen into Modern Times.* Hodder and Stoughton (London).
- Watson, J.D. and Crick, F.H.C (1953). A structure for deoxyribose nucleic acid. *Nature,* 171: 737–738.

A PRELIMINARY, COMPARATIVE TYPE PROPOSAL FOR LARGE, UNIDENTIFIED MARINE AND FRESHWATER ANIMALS

Bruce A. Champagne
Email: bruce.champagne@lycos.com

ABSTRACT
In a previous work proposing unidentified marine animal types based on a database analysis of observations, the author suggested a comparison of proposed, scientifically undescribed marine animals to unidentified lake animals may advance the data set of each (Champagne, 2001). After reviewing the generic descriptions of 163 unidentified freshwater, or suspected euryhaline aquatic animals, the author suggests that forms of more frequently observed, although scientifically unclassified, freshwater animals may be migratory, transient, or resident variants of the author's marine proposal, the Type III Multiple Humped Animal. The author suggests the Type III Multiple Humped Freshwater Animal proposal is falsifiable and can be utilised and compared to subsequent observations and a planned observation database analysis, and potentially maximise limited observational opportunities and/or assist with a type specimen collection.

Keywords: sea serpents, lake monsters, cryptozoology

INTRODUCTION

The author preliminarily reviewed 163 independent, summarised descriptions of unknown freshwater animals, sometimes referred to "lake monsters", and six descriptions of animals observed in both marine and freshwater environs throughout the world. The author noted that two types of his previously-proposed unidentified marine animal types appeared similar to some unknown freshwater and inland aquatic animals. The apparent similarities not only initially extended to possible morphology, but also to potential behavioural and geographical factors.

The author was not the first to suggest a possible correlation between currently unknown marine and freshwater animals, sometimes referred to as "sea serpents" and "lake monsters" respectively. Other researchers have also commented on a possible similarity (Costello, 1974; LeBlond and Bousfield, 1995), perhaps featuring a creature similar to Heuvelmans's long necked pinniped proposal (1968) (Fig. 1).

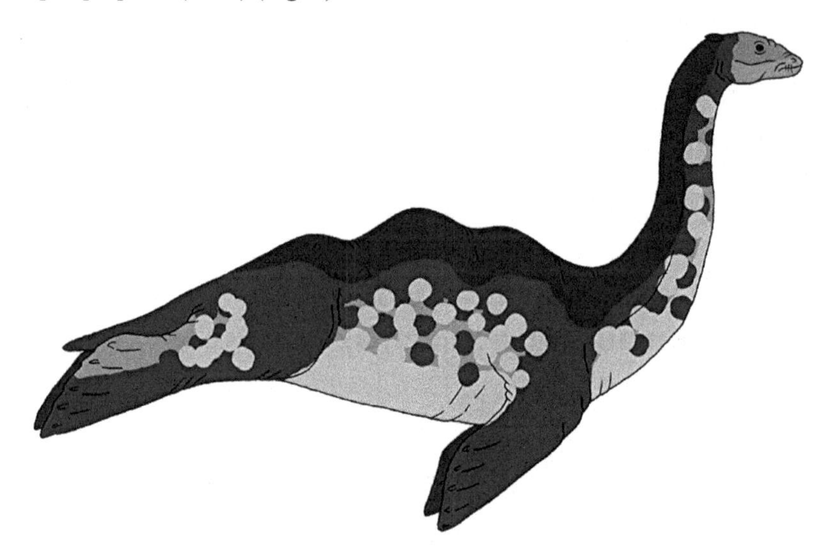

Fig. 1: Artistic representation of Heuvelmans's proposed long-necked seal (© Tim Morris)

Oudemans (1892), and later Costello (1974), suggested the unidentified animal in Scotland's Loch Ness, informally referred to as "Nessie", was a freshwater form of long necked/long-tailed seal. Olsson (1899) suggested a similar proposal for the animal observed in Sweden's Lake Storsjön, with Rafinesque-Schmaltz (1819) suggesting a connection between sea snakes and the unidentified animal reported in North America's Lake Erie. Although Harrison (2001) believes that these currently unknown marine and freshwater forms should not be combined when referencing unidentified lake animals in the British Isles, Shuker (1996) suggested some unidentified freshwater animals may be similar to the marine types of the same general area but are not likely the same species, although presumably closely related.

Coleman and Huyghe (2003) opined that there have been no serious attempts to classify the lake monsters reported around the world because, in part, they are believed to be of a singular nature or type - some form of "long-necked" animal. They also suggest that the majority of any scientific interest appeared to revolve around "piecemeal" attempts to identify the animals. Encouragingly, LeBlond and Bousfield (1995) acknowledged that some of the unidentified marine animal reports that they examined suggested some resemblance to lake forms, and proposed that the study of unidentified freshwater animals may profit from data associated with the oceanic types.

Common arguments in support of or against the existence of relatively large, scientifically undescribed freshwater animals may be typically based on the presumption that the animals are apex predators, and on the population of, and size and metabolic relationships to, proposed required standing fish stocks as primary food sources (Sheldon and Kerr, 1972; Scheider and Wallace, 1973; Mangiacopra, 1992), or estimations of observational effort. Other, sometimes contradictory arguments may also be founded on a specific freshwater body's size, climate, geographical history, and/or access to the ocean. Although Coleman and Huyghe (2003) suggested that the longest traditions of potential lake monsters were from larger lakes, Harrison (2001) reported that he found no correlation between Scottish loch size and reports of reported lake monsters, whereas Radford and Nickell (2006) suggested that lake monster sightings merely corresponded to public interest.

GENERAL "LAKE MONSTER" REVIEW OF DISTRIBUTION AND DESCRIPTION

The author examined general descriptions of the more notable "lake monsters" where there appears to be reasonable agreement of morphology and behaviour.

It is probable that some unidentified lake animal reports have been hoaxes, misidentifications, and over-simplified descriptions of unknown, and sometimes legendary or mythological freshwater animals. Descriptions have included a long, cylindrical neck and/or a plesiosaurid body plan, or even a composite of this romantic long-necked type and the horizontally-oriented, lobed tail frequently reported with a serpentine form (Fig. 2). The author noted a similar condition when examining unknown marine serpents, and theorised that some descriptions were likely the same type of animal (Champagne, 2007).

Fig. 2: Type 1A Long-Necked Animal (© Tim Morris)

Other researchers have also made a type distinction between various freshwater "monsters". Some of those examinations were based solely on behaviour (Kojo, 1992), whereas others were based on morphology and preferred environments (Watson, 2011).

There has also been some previous estimation of the distribution and potential habitat of these unidentified freshwater animals. Reporting on approximately one thousand alleged lake monster locations, Coleman (2001) also quoted Bernard Heuvelmans's observation: "Attention must drawn to the fact that all these long-necked animals [so-called "Lake Monsters"] have been reported from stretches of freshwater located at isothermic lines 10°C, that is, between 0°C and 20°C in both Northern and Southern hemispheres".

Mangiacopra (1992) identified the basic morphology, probabilities for existence, and potential population densities for some of the more frequently reported, unidentified freshwater animals in North American lakes. He found that the unidentified animal referred to as "Memphre", purported to inhabit Lake Memphremagog in Vermont and Quebec, Canada, has been described most frequently as a 7.6-9.4-m serpent with multiple humps, and secondarily with a horse-like head on an elongate, snake-like fish. Although Mangiacopra was unable to estimate a population size, he allowed for the animal's possible egress or migration or both from the lake.

North America's Lake Champlain is said to be the home of a scientifically-undescribed animal generically referred to as "Champ". Although the animal has been compared to Heuvelmans's marine "Super Otter" (Coleman, 2009) and a recent form of the middle Triassic, long-necked reptile *Tanystropheus longobardicus* (Hall, 2000; Elizabeth and Hall, 2014) (Fig. 3), the majority of sightings appear to describe a 4.6-9.2-m-long serpentine animal swimming, often perceptually very quickly, with 1-10 vertical undulations and humps (Kojo, 1991) and/or a snake-like or round head with two small head projections.

Fig. 3: Life-sized model of *Tanystropheus longobardicus* (© Karl P.N. Shuker)

Mardis (2015) has proposed that some Champ observations suggest the presence of not only the serpentine form but also a long necked form with a more plesiosaurid body plan. Unless Champ is capable of overland locomotion around two drainage obstructions (Kojo, 1991), Mangiacopra suggested this animal is the only lake monster candidate that could potentially be considered a homeotherm, with a population of up to ten individuals. Possibly supportive of a modern or relict cetacean proposal, von Muggenthaler recorded what she believes were echolocations in Lake Champlain in 2003 (von Muggenthaler *et al.*, 2010), although there is some disagreement that the echolocations are consistent with known cetaceans in pattern and cadence (Bartholomew, 2012).

Elizabeth (2014) has maintained the *Tanystropheus* (or analogue) proposal as being capable of echolocation as an evolutionary adaptation after recording possible echolocations in 2014. Based on an examination of observational distributions, Kojo (1992) suggested Champ to be primarily nocturnal and possibly dissimilar from other prominent lake "monsters".

"Ogopogo" reportedly inhabits Lake Okanagan in British Columbia, Canada. Generally, this unknown 6-12-m-long serpentine animal is described with 2-6 humps, a horse-like head, and a forked/fluked tail (Eberhart, 2002). As observational distributions are different from the proposed animal in Lake Champlain, and more similar to the animal in Loch Ness, Scotland, Kojo (1992) suggested Ogopogo to be diurnal and a different species from that reported in Lake Champlain. Mangiacopra (1992) concluded that if Ogopogo were mammalian, only three individuals could be environmentally supported as residents in the lake. Although Mackal (1980) proposed that Ogopogo may be a relict basilosaurid whale (Fig. 4), Bousfield suggested the animal was related to his proposed marine type, *Cadborosaurus willsi* (Eberhart, 2002).

Fig. 4: Artistic reconstruction of a basilosaurid whale (© Tim Morris)

Frizzell (2001) made a size distinction between "Chessie" of the Chesapeake Bay area of North America, and the marine animal of the proximate coastal area, sometimes referred to as the "Great New England Sea Serpent" (O'Neill, 1999). Generically described as a 4-8-m-long serpentine animal, Chessie displays 3-4 humps, a turtle-like head, a 3-m-long neck, and has been reported swimming with both horizontal and vertical undulations (Eberhart, 2002). Frizzell (2001) also recounted reports of an unknown animal in the adjacent Choptank River and stories from watermen living in the swamps off Nanjemoy Creek of the animal "slithering through murky waters".

Lake Erie is the reported habitat of "South Bay Bessie", a 10-18-m-long serpentine animal with a long neck, 3-5 humps, and a flat tail (Eberhart, 2002). "Cressie", the unidentified animal of Crescent Lake in Newfoundland, has been identified as a 6-m eel-like animal and occasionally described with a horse-like head. Named after the river of southeastern Georgia in which it has been observed, the "Altamaha-Ha" is reported as a serpentine animal, 3-5 m in length, with a small head, long neck, and 2-3 humps displayed (Eberhart, 2002).

Other unknown North American animals have included observations of an unidentified animal entering the estuarine St. Croix River in Maine with the incoming tide, an unidentified animal in the St. Johns River of Florida, and the multiple-humped animal observed in the Sacramento River of California, approximately 113 km inland from the San Francisco Bay (Champagne, 2007).

The United Kingdom is a historic region for various, and at times celebrated, undescribed aquatic animals. Watson (2011) differentiated between the legendary "water horses" associated with lakes, and the "kelpie", reportedly found in rivers. Generally, the serpentine, equine-headed water horse has been reported in the proximate ocean, lakes, rivers, and streams of the British Isles, as well as Scandinavia and elsewhere in mainland Europe, whereas the horse-headed kelpie is reported in both Scottish lakes and rivers, and particularly in swift streams (Eberhart, 2002).

Prominently, "Nessie" of Loch Ness, Scotland, has been described as a long-necked, humped animal with a head often indistinguishable from the neck, and, alternatively, as a serpentine animal with a head equine in appearance. Although Bauer (2002b) noted in his examination of the 1960 Dinsdale film evidence of the alleged animal paddling with limbs/fins at the sides, multiple other observations have reported locomotion produced by the tail, which, perhaps supportive of the author's proposal, has also previously been reported as bilobate (Costello, 1974). Kojo (1992) proposed Nessie to be a diurnal animal based on an analysis of observation time distribution, whereas Watson (2012) suggested Nessie may be more likely to venture onto land nocturnally.

Additionally, there are the humped, fork-tailed animals of Lough Fadda, Ireland, a 10-m-long, multiple-humped serpentine animal observed in Loch Morar, Scotland, the serpentine animal confined to a culvert proximate to, and between, Lough Crolan and Lough Derrylea, Ireland, the unknown animal trapped under a bridge in a stream in Ballynahanich, Ireland until rain and rising water levels helped the animal to escape, and the 3-m-long animal crossing a roadway with "swaying" undulations by Sraheens' Lough, Ireland (Champagne, 2007).

Other animals similar to those referenced include undescribed, freshwater animals reported in continental Europe (particularly Scandinavia), Africa, southeast Asia, Russia, Japan, Australia, and South America.

RESULTS

The author examined the generic descriptions of 163 unidentified freshwater, or suspected euryhaline, aquatic animals. Summaries of the data amassed by him are presented in this paper's Appendix. Approximately 38% of the descriptions estimated the size of the unidentified animal. Reported, general sizes overlapped in class with 3% of all sightings of an unidentified freshwater animal describing a size less than 3 m total body length (TBL). 26% of all sightings included a reported size of 8-18 m TBL, with 9% of sightings estimating a size in excess of 16 m TBL.

When colouration was specifically reported, 18% of the freshwater animals were described with a dark colouration and 3% with a yellow-orange colouration. When humps were reported, 4% of the types reported only one hump, whereas 18% of the animal types reported more than one visible hump.

Serpentine animals were described in 28% of the freshwater types and with a long-necked description in 15% of the animal types. Specifically described head morphology consisted of 21% of the types reporting a cameloid/equine head (it should be noted here that Paxton, 2010, observed that grey seals *Halichoerus grypus* in the Canadian Maritime provinces were also called "horseheads"). Plated, horizontally-oriented bilobate tails were reported in 8% of the descriptions, a sirenian/pinniped-like body in 8%, a fish-like body in 3%, a crocodilian body in 2%, a large "bulky" body in 2%, and a chelonian-like body in 1% of the types. Flippers were reported in 5% of the types, and legs described in 5% of the freshwater observations. The percentages of the total number of observations of each type of freshwater animal appear similar to the corresponding, or similar, marine types (Champagne, 2007).

The undescribed animals reported from regions and bodies of freshwater, and estimated to be most likely to exist based on previous research and environmental requirements, were compared by description, location type, and seasonal observations. Descriptions included the head, body, and tail. Locations were based on latitude, lake or river environment, direct/indirect access to the ocean, and seasons of observations. Although some of the previous, general proposals of unidentified freshwater animals may be of relict marine or modern freshwater reptiles, species possibly convergent with a plesiosaurid body plan (Mardis, 2014), fish, cetacean, or long-necked pinniped, each of the generic descriptions of the undescribed freshwater animals compared favorably to the author's marine Type III Multiple Humped Animal proposal (Fig. 5).

Animal	Head	Body	Tail	Size	Latitude	Location	Environment	Season	Time	Ocean Access
Champ	Equine/Snakelike	Serpentine	Fish-like	4-10 m	43-45	Lake	Oligotrophic	Summer	Late PM	Obstructed
Kelpie	Equine	Varied	Varied		55-56	Rivers/Lakes	Varied			Varied
Memphre	Equine	Serpentine		7-10 m	45	Lake	Mesotrophic	Summer		Yes
Nessie	Equine	Serpentine	Bi-lobate	3-7 m	57	Lake	Oligotrophic	Summer	AM/PM	Obstructed
Ogopogo	Equine	Serpentine	Forked	6-23 m	47	Lake	Oligotrophic	Summer	AM/PM	Obstructed
Water Horse	Equine	Serpentine	Whale-like	3-7 m	55-56	Lake/Rivers	Varied	Fall		Varied
Type III	Equine/Cameloid	Serpentine	Horizontally Bi-lobate	3-20 m	40-50	Ocean, Estuary	Marine, Brackish	Summer	AM/PM	Yes
Type I	Reptilian	Plesiosaurid	Singular	3-12 m	30-40	Open Ocean	Marine	Summer	AM	N/A

Fig. 5: Comparative Matrix of Prominent Unidentified Freshwater Animals

All of the compared, representative undescribed freshwater animals were in 100% agreement with the equine, "horse-like" head description of the Type III Multiple Humped Animal proposal. With the exception of the varied body description of the Kelpie, all the compared animals were also in agreement with the serpentine body description of the Type III Multiple Humped Animal, although the animal reported to reside in Lake Memphremagog does not have a report providing a tail description and the Kelpie's tail description is varied. The remaining compared animals were in 100% agreement with the distinctive, horizontally lobed tail and all size estimates were within the size ranges of the Type III Multiple Humped Animal. The representative animals also appear to occupy the same ranges as the Type III Multiple Humped Animal and, with the exception of the autumn movements and observations of the Water Horse, each animal was more likely to be observed in the summer, also in agreement with the Type III Multiple Humped Animal.

The author's generic marine Type I Long Necked Animal proposal was also compared to the generic, unidentified lake animal descriptions. The Type I Long Necked Animal proposal was in agreement in general size with the ranges of Champ, Memphre, Nessie, and the Water Horse, and in partial agreement with the size ranges of Ogopogo. The Type I Long Necked Animals' ranges overlapped Champ's and was temporally distributed similar to Nessie and Ogopogo.

UNIDENTIFED MARINE ANIMAL STUDY REVIEW

In review, the author collected and analysed observational data from over 118 sources to include personal communications, texts, journal articles, field reports, television/news accounts, newspapers, and the worldwide web, resulting in 1247 collected observations of unidentified marine animals. These observations were entered into a spreadsheet for organisation and sorting, receiving an observation tracking number that included the reference source of the report (see Appendix for summaries). Available and extrapolated data of each observation were recorded in a number of areas for analysis.

An observation then received a report description number unique to that observation. The description number includes the assigned observation number and a letter designation indicating particular elements of the observation. The description number includes the following data if appropriate to the observation: a designation for an actual observation of an organism (as opposed to a wave phenomenon, or water disturbance), a behaviour of note exhibited by the animal, an act of predation observed or implied by the observer, a designation of an observation of a carcase or moribund animal, an observation where the locomotion of the animal was described, an observation by a qualified person (life scientist, or individual familiar with local fauna, etc), vocalisations emitted by an observed animal, an observation of a juvenile animal (as judged by a substantial size difference when in the presence of another animal of the same type, or when compared to the average size of a similar animal), an observation of multiple animals during the same observation, the initial response (interest, fear, etc) of the observer(s), a physical characteristic of note observed on the animal, aggression displayed by the observed animal (either towards other animals or towards the observer), an observation recorded either before or soon after a natural event (earthquake, flood, storm, etc), an observation directly related to a recent event (natural or man-made), an observation recorded in a location of note, and a designation of a capture or photograph (or both) of an observed animal.

The date, time of day, and season (if the date is incomplete) were also entered into the spreadsheet. Seasons in the northern latitudes were designated as Spring: 20 March to 19 June; Summer: 20 June to 19 September; Autumn: 20 September to 19 December; and Winter: 20 December to 19 March. In the southern latitudes, seasons were designated as Spring: 20 September to 19 December; Summer: 20 December to 19 March; Autumn: 20 March to 19 June; and Winter: 20 June to 19 September.

A rating was assigned and recorded for each observation. That rating reflected the quality and accuracy, as determined by the elements of the observation rated for credibility with predetermined criteria (see Quality Control of Data).

The duration, or length of the observation, in addition to the distance at which the animal was observed, was recorded. Longitude and latitude coordinates were estimated from information provided in the reports and recorded. Physical characteristics of the water conditions and parameters from the location of the observation were recorded using general information provided by various texts, unless specifically stated by the observer. These characteristics included depth (National Geographic, 1990), salinity, surface temperature, water temperature (Ingmanson and Wallace, 1985), typical wave and tidal conditions and characteristics, and the predominant current (Couper, 1983), if applicable. Additionally, atmospheric pressure, sea floor morphology, relation to earthquake frequency depth ranges, submarine heat flow, submarine sediment distribution, predominant winds, average wind speeds, mean annual rainfall, water column movement, typical thunderstorm activity, marine biogeographical area, phytoplankton production in the area, zooplankton production in the area, and benthic biomass production in the area, as provided by Couper (1983), the weather conditions at the time of the observation, and the weather conditions three days prior and three days subsequent to the observation were also noted in the dataset, if provided by the observer.

The number of observers per report, their activity at the time of the observation, and their immediate reaction to the observation were also noted. So too was the number of individual animals per observation, in addition to the numerical value assigned to the general type of animal observed.

Initially, an observation was classified with at least one of the following designations/ descriptions derived from Heuvelmans's 1968 list (simplified for brief observations) of unidentified aquatic animals. It would become evident that some of Heuvelmans's animal types were not specific enough for differentiation, essentially allowing for the inclusion of a reported animal into more than one type description. A second evaluation of the observations was then conducted, and additional, more specific descriptions of animal types or variations (or both) were included. The final animal types included the Long Necked Animal, Eel-Like Animal, Multiple-Humped Animal, Unidentified Animal (utilised when morphological details were unavailable and/or insufficient for inclusion into another animal type grouping), Other Type of Animal, Dorsal-finned Animal, Carapaced Animal, Cephalopod, Lizard-like Animal, Multiple-Limbed/Segmented Animal, Digited Animal, and Scaled Snake-like Animal. After the descriptions were determined, the labels were then designated as Type I, Type II, Type III, etc. The physical dimensions of the animal were recorded, in addition to the animal's activity when initially observed, and the animal's response to the observer.

When possible, behaviours, descriptions, and locations were assigned numerical values to aid in the sorting process.

When the complete dataset was recorded (1247 observations), it was sorted by a numerical, credibility rating. Repetition was also checked to ensure that no observation was included in the dataset more than once. The observations were sorted by credibility rating (>=5), with observations under the credibility rating discarded from the final data set and analysis.

The sorted dataset was then graphed (histogram, etc) for each category and compared to current proposals, by referencing the data to existing literature. Patterns and behavioural trends were then established for each type of animal, and, subsequently, patterns specific to a particular geography.

After the evaluation was completed, an abbreviated natural history was developed for the animal(s) types identified. A selected extant species, or, where more appropriate, a species from the fossil record, was examined and compared to increase the accuracy of the proposed natural histories. Comparison with an extinct species was necessary in order to address the theories of other researchers maintaining that the animals may be relict organisms (Mackal, 1980).

QUALITY CONTROL OF DATA

As collected observations may have been old, second or third hand accounts, of questionable authenticity, or a case of mistaken identification, a rating scale was developed to evaluate and estimate the credibility and usefulness of each observation. It was understood that no evaluation process would be error-free when reviewing all observations. In addition to unanticipated complications, some observations undoubtedly are complete or partial fabrications. Also, the observer may have been sincere in relating the observation, but mistook a type of wave phenomenon or a known species or object for the unknown animal. It was also considered that a false identification may be the result of distance (too great for accuracy, detail, or proper identification), lighting, emotion (fear, etc), duration of the observation (too short for a detailed observation - Andrews, 1935), and/or lack of any qualification (education, experience, etc), to evaluate and describe objectively and accurately what was observed.

Each observation was reviewed and awarded points for an element met from a credibility rating list. A maximum score of eleven points could be awarded for each observation.

One point was awarded for an observation that was reviewed by a competent investigator and found to be credible. The investigator would have been experienced and objective in such observations, and was educated in a life science discipline.

A point was awarded for an observation made by a qualified observer. An observer was considered qualified by being formally educated in a life science, experienced through occupation (i.e. commercial fisherman), or with above-average observational skills.

If the observer provided a detailed account or physical description of the animal, location, animal's activity, or elements of the account (date, duration, weather conditions, etc), the observation was awarded one point. Descriptions such as "a big, snake-like thing" or "it looked like an upturned boat" were not considered detailed descriptions.

An observer with a known reputation for veracity, or with an occupational or societal position that may suffer or be lost if the observer were linked to a fabrication, was also taken into account. Consequently, the observation was awarded one point for the potential credibility of the observer(s).

If an observation was documented as simultaneously witnessed by more than one individual, it was awarded one point. It was assumed that if more than one observer witnessed the event, the opportunity for exaggeration or fabrication would have been reduced.

Through personal field experience with brief observations (and later verified by video) of similar sized animals (great white shark *Carcharodon carcharias,* Californian sea-lion *Zalophis californianus,* northern elephant seal *Mirounga angustirostris,* Steller's sea-lion *Eumatopias jubatus,* harbour (common) seal *Phoca vitulina,* killer whale *Orcinus orca,* blue whale *Balaenoptera musculus,* grey whale *Eschrichtius robustus*, etc) made at a distance on the surface by the author, it appears that there could be a maximum distance that an observer may provide a valuable, accurate narrative with the unaided eye. The author's personal experience provided an effective observational distance with unaided eyesight, at approximately 62 m. An observation made at or under this distance was awarded one point.

The longer the observation, the more accurate and detailed an observation could be expected. The likelihood of misidentification may also be reduced. An observation of 60 seconds or more was awarded one point.

If an observation included tangible, physical evidence (carcase, tissue samples), it was awarded two points.

A photograph, film, video, or sonar reading helped to substantiate some observations. As a result, an observation that was documented by photographic evidence, verifiable sonar reading, or both was allowed one point.

Biogenic structures (changed landscape, prints, damage to involved objects, etc) further substantiated an observation. Observations with such corroborating data were awarded one point.

A value of five points was initially decided as the minimum standard for an observation to be included in the final data set. By using this benchmark, it was assumed that an observation would have more probably occurred. All observations received one point. By design, most observations fell under the tolerance of five points and were excluded from the final data set. It is understood that even if a point was awarded for a particular category, this award does not necessarily indicate that the category's strength was individually sufficient to substantiate an observation. This point system was designed to be used in totality, as no single category was intended to stand alone as corroboration.

QUALITY CONTROL RESULTS

After the sort by credibility rating, 358 observations (29%) remained from the original data set of 1247 observations. There was one observation with a credibility rating of eleven points. There were no observations with ten points, and four observations with a credibility rating of nine points. Fourteen observations were recorded with eight points, and thirty-eight with seven points. Eighty-two observations were awarded six points, and 209 received the minimum of five credibility points.

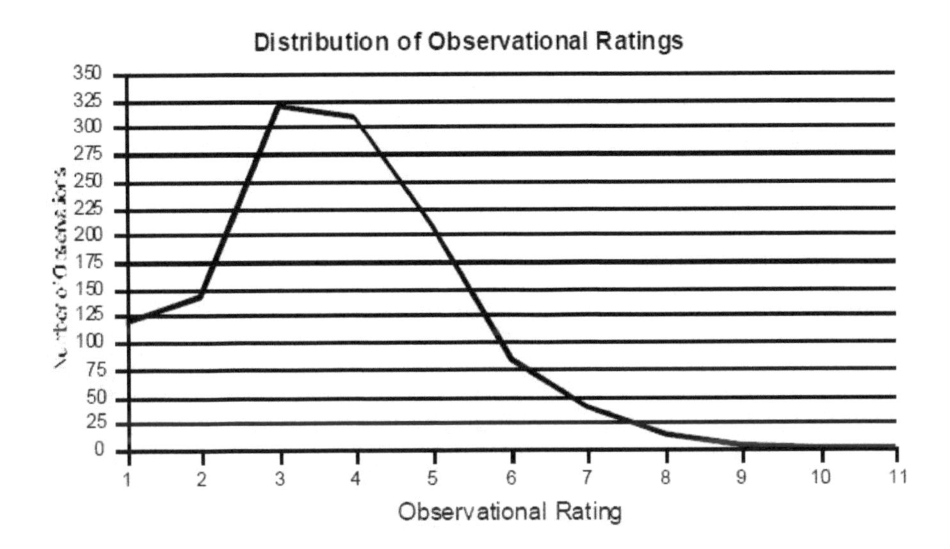

Graph 1: Distribution of Observational Ratings

The distribution of all observations (N=1237) by credibility rating score, with a mean score of 3.686, and a standard deviation of 114.782

Of those observations discounted from the data set, 309 received four points, 320 received three points, 142 received two points, and 118 received one credibility point.

Initially, the minimum credibility rating score was theorised to be five points. It was presumed that an observation with this minimum score would more than likely have taken place with a reduced opportunity for deception, misidentification, or both. Statistical analyses were completed from the observational data. To assist with subsequent research and comparison, t tests, z scores, and a Level of Significance One-Tailed Test were completed to evaluate the minimum retention score, and sensitivity of five.

Observations with a credibility rating of one did not meet the significance criteria. Observations with a credibility rating of two were significant to .0005. Observations with credibility ratings of three and four were not significant. Observations with a credibility rating of five were significant to .001, with observational credibility ratings of six, seven, eight, and nine significant to .0005. There were no observations with a credibility rating of ten, and one observation with a credibility rating of eleven, with an associated significance to .005.

A consistent, observational threshold appears to begin with observations rated for credibility with a score of five points, and in agreement with the original theory of a minimum standard of five points.

TYPE III MULTIPLE-HUMPED ANIMAL DESCRIPTION

The initial Type III Multiple Humped Animal marine description was isolated from other observations by the report of two or more humps observed at the surface. Subsequently, more specific characteristics were included and proposed as Type III Multiple Humped Animal observations eventually constituted 29% of all marine sea serpent reports in the author's original study (Fig. 6).

Fig. 6: Type III Multiple Humped Animal (© Tim Morris)

Using a classification matrix, the Type III Multiple Humped Animal scored between plesiosaurs and primitive mammals. The Type III Multiple Humped Animal appears to attain a maximum length of 60 m TBL, with the size class of 10-20 m TBL most commonly reported. The diameter of the body is 3-10% of the TBL and cryptically coloured, possibly indicating adaptation to foraging along a dark substratum while minimising a presence to potential predators and/or absorbing the available solar heat at the surface. The skin is smooth in appearance and texture and may also possess an erectile dorsal spine arrangement. Depending upon the viewing angle of the observer(s) and the available light, small spines (often described as yellow in appearance) may appear as a hair-like skin. It is possible that the Type III Multiple Humped Animal uses this skin system to gain a purchase and maintain its position on slippery rocks, substrate, or both during feeding in a wave-swept environment or while basking on land.

There are two pectoral fins or appendages positioned approximately 30% of the TBL from the head. A dorsal crest or scutes are present, and a mane has been observed on occasion and on animals as small as 3.0 m TBL.

The distinctive bilobate tail is horizontal in orientation, and "plated" along the dorsal surface. It appears the tail may be retracted or withdrawn to create a minimal profile, thus reducing drag in selected environments and situations, or on the recovery stroke of the swimming motion, or both. The tail's appearance, perhaps because it superficially resembles a crustacean telson, has been described as "shrimp-like" or "lobster-like". This rigid tail structure may be used as a support for the soft-rayed tail during the power stroke of the tail's swimming motion, when gaining a purchase on substrate for locomotion, or both. It appears that the Type III Multiple Humped Animal may also use the tail structure to generate noise. One observation described the rattling sound produced by the tail of a startled, basking animal like that of a "Gatling gun", with an additional observation referencing a "vibration" associated with the unknown animal (Champagne, 2007). Perhaps, the animal employs the vibrating or "rattling" tail to warn potential rivals or predators of the animal's presence, to demonstrate aggression, or both.

The equine or cameloid head length dimensions appear to be 5-7% TBL, with the width of the head at 3% of the TBL. Ears or horns have been observed on occasion. The eyes have been described as green in colour, and forward facing, suggesting binocular vision and a predatory strategy. The mouth is sub-terminal to ventral in position, suggesting that the Type III Multiple Humped Animal exploits feeding opportunities primarily at or near the sea floor (Nikolsky, 1978). The teeth have been described as fish-like and sharp with a palate that appears to be textured, possibly serving as an organ for crushing or securing prey in the mouth (Storer *et al.*, 1979), an adaptation for vomerine teeth, or a natrial gland for salt excretion/ regulation. A tongue has been observed, and was described as long, rigid, and "spear-like". Perhaps the structure of the tongue allows the Type III Multiple Humped Animal to impale and retrieve prey from small crevices. Large, prehensile lips have also been reported. These lips may aid the animal in browsing and removing plant material or small food items from crevices and difficult-to-access substrata by providing a tactile and/or suction advantage. The aerial senses of the Type III Multiple Humped Animal appear to be limited, but observations suggesting enhanced, sub-surface visual and auditory senses have been recorded.

The Type III Multiple Humped Animal is likely to be solitary in nature, and may only be found in the presence of conspecifics during breeding season. It has also demonstrated a form of territoriality or aggression towards other species (*Zalophus californianus*), possibly as potential or perceived rivals during mating, or competitors to the available food supply.

The Type III Multiple Humped Animal has been identified with 2-30 humps at the surface of the water, with three being the most frequently reported number, and it appears capable of swimming in excess of 10 km/h. When the animal is perceived to be swimming rapidly at the surface, it is always with a humped posture and vertical undulations. The hump directly behind the head has been described as stationary with the humps following it apparently undulating. LeBlond and Bousfield (1995) proposed that *Cadborosaurus willsi*, a similar animal reported from the waters of Cadboro Bay on Vancouver Island, western Canada, assumed a humped position to approximate more closely the body shape of a fast-swimming thunniform, therefore assuming a more efficient swimming body plan. When the animal is observed basking, floating motionless or swimming slowly, its body is usually extended.

It appears that the Type III Multiple Humped Animal is behaviourally oriented towards the air-water interface, with head displays (spy-hopping [SH] and periscope behaviour [PB]) constituting 49% of all recorded observations. It is possible that these behavioural displays, in the absence of visible humps, have caused the Type III Multiple Humped Animal to have been misidentified as a Type I Animal (Long Necked Animal) on occasion (Fig. 7).

Fig. 7: Type IB Long Necked Animal (© Tim Morris)

Type III Animals appear to migrate daily towards the shore or inland waterways after 00:00 hours and under the cover of darkness. The animal then returns off-shore, possibly into the open ocean, before 06:00 hours.

Although the Type III Multiple Humped Animal has been observed in both hemispheres, it is more frequently reported in the northern hemisphere (90% of observations), with a preferred range of 40-50° latitude. The data suggests this type's preference for the temperate waters over and proximate to the continental shelf, and the boreal air temperatures of shallow and in-shore waterways.

A seasonal migration has been identified within the Type III Multiple Humped Animal's north-south range. During the summer months, the animal appears to migrate as far as 70° latitude, only to return to proximal equatorial latitudes during the fall and winter months.

The Type III Multiple Humped Animal may give birth to precocial young in the relative protection of inland waterways. Animals observed inland are from the extremes of the larger and smaller size classes (4-30 m TBL). With the observation of smaller size-class animals in the higher latitude ranges, it was proposed that a Type III Animal may mate and give birth in these more fertile areas in order to maximise the growth rate and survival of offspring, returning to the lower latitude range only as adults.

In 63% of the total recorded observations, the Type III Multiple Humped Animal was reported in proximity to estuaries or river confluences or both. The majority of observations also included geographical or environmental conditions with a relative nearness to fully or semi-protected shorelines, multiple islands, offshore banks, and/or strong tidal currents. In areas where the Type III Multiple Humped Animal has been reported with greatest frequency, large and productive estuaries, or river confluences, or all of these features were in the immediate vicinity. The animals observed more than 5 km inland, were at times observed on mud flats and basking on exposed banks.

The Type III Multiple Humped Animal has been reported in fresh or brackish water, and appears to have a preference for seawater of less than 33% salinity, likely venturing into estuaries, up rivers, or both, as observed with the large animal in observation #259 of the author's study of undescribed marine animals (Champagne, 2007). There are more observations of the Type III Multiple Humped Animal swimming up rivers and tidal areas against the outgoing tide, as opposed to movements with tidal movements. Perhaps the Type III Multiple Humped Animal exploits the more concentrated prey organisms or the more exposed prey organisms (or both) for feeding opportunities offered by the lower water levels of the receding tides. The Type III Multiple Humped Animal may also have the capability to move over land for short distances. Alternatively – or additionally - it may not hesitate to negotiate marshes when conditions permit.

Typically, the areas of the reported observations were of a high (>250-500 $mgC/m^2/d$) phytoplankton production, moderate to high (>50-500 g/m^2) zooplankton production, and moderate (1.0-300 g/m^2) benthic biomass production. When the mouth position and morphology of the Type III Multiple Humped Animal and available food were taken into consideration, it was proposed that a Type III Animal likely preys upon moderately-sized species of zooplankton and benthic bony fishes and invertebrates. Although the Type III Multiple Humped Animal has been recorded preying upon seabirds, it is suggested that surface species may not be the usual or preferred prey items for the animal. Such events may be an automatic predatory strike, or the exploitation of a seasonally-available food source.

The Type III Multiple Humped Animal appears to be comparable to an animal observed and recorded from many locations throughout the world (Eberhart, 2002). When a comparison of some of the more frequently reported and similar animals to the Type III Animal is made, it appears that the animals are likely members of the same genus and/or species (*Cadborosaurus*, the Great New England Sea Serpent, U.K., and Indochina).

TYPE III MULTIPLE HUMPED FRESHWATER ANIMAL

Although more than one type of unidentified animal could be the source of lake monster sightings and legends, the author suggests that the most likely candidate for unidentified inland freshwater animals is the Type III Multiple Humped Animal or variant. Theorised and observed behaviours, distributions, environmental requirements, and reported descriptions of the marine Type III Multiple Humped Animal are supportive and in agreement with the elements of multiple freshwater reports, historical artifacts, and legends. Resident, migrant, and transient variations of this proposed animal may be compared to not only the more noted freshwater bodies, but also the less frequently reported lakes, rivers and streams, bogs, and swamps. Additionally, the author suggests that this type proposal may answer the varied arguments suggested against (or limiting) the existence of lake monsters generally.

Among the arguments against or as a restriction of types are those of a prohibitive climate, water temperature, or both. Although more than one researcher, including the author, has suggested that the Type III Multiple Humped Animal or variant is reptilian, the colder climates of some environments and water would presumably be environmental barriers to the existence or range of this type of animal, even though this particular animal appears to be most frequently reported in these areas, based upon the provided descriptions and behaviours (Champagne, 2007).

However, both the fossil record and known, extant species appear to provide precedents for aquatic reptile tolerance of cooler or cold water temperatures and environmental conditions. After examining polycotylid plesiosaur fossils from non-marine deposits in the Hauterivian-Aptian Formation, Vavrek *et al.* (2014) suggested that not only did these animals, of various size classes and age groups, inhabit high latitudes with seasonally cool temperatures, but also they may have ventured periodically into freshwater environments. Kear (2005) examined the fossils of three families of plesiosaur (Elasmosauridae, Polycotylidae, and Pliosauridae) and possibly one family of ichthyosaur from the White Cliffs deposits of southeastern Australia, and found that these fossils were located in association with palaeoclimatic indicators suggesting seasonal, very cold to near freezing conditions.

Seymore (2004) suggested current evidence supports a hypothesis that ancestors of crocodilians were endothermic, based on their four-chambered heart, thus providing the ancient reptiles with the physiological capabilities associated with the metabolic rates and temperature tolerances of endotherms. Another function of the associated, high systemic blood pressure is support to an extended, vertical blood column above the heart. Such a capability would not only allow a physiologically similar animal to exploit temperate to cold environments, but also provide the required support for the reported spy hopping and periscope behaviours demonstrated by the Type III Multiple Humped Animal. Other recent and supportive research has also suggested that ancient marine reptiles were capable of maintaining high body temperatures independent of the water temperature, even in cold and temperate environments (Bernard *et al.*, 2010).

Perhaps a Type III Animal is endothermic, with the ability to hold and regulate a thermal gradient (Bostron, *et al.*, 2010), or is gigantothermic, with a body size sufficiently large for internal temperature to be minimally affected by the environment. The Type III Multiple Humped Freshwater Animal may have a temperature tolerance similar to other aquatic reptiles. American alligators *Alligator mississippiensis* have been monitored, with no apparent physical distress, in temperatures as low as 7°C (Colbert *et al.*, 1946), and are known to display an "icing response" when their aquatic environment freezes and the animal is essentially contained under the surface ice, exposing only its nostrils above the frozen surface (Brisbin, 1982). Also cold tolerant, the painted turtle *Chrysemys picta* is known to hibernate in the mud at the bottom of a pond during the winter, and may even emerge to bask on some warmer winter days (Ernst and Lovich, 2009).

Some freshwater bodies reportedly containing unidentified animals are covered with surface ice during the winter. Such a challenging environmental condition would presumably also prevent an air-breathing animal from permanent residence in the lake. Perhaps because a Type III Animal is possibly capable of overland travel, the animal may not remain in a freezing lake or river. Instead, the animal may move to another body of water or a coastal environment, or remain on land with minimal metabolic activity. Alternatively, the lake animals may remain in a frozen body of water, breathing from specialised adaptations (Mardis, 2014) or available air spaces under the ice, perhaps similar to the suggestions Smith and Horonowitsch (1987) made regarding harbour seals *Phoca vitulina* in the Hudson Bay drainage, or the "icing response" of American alligators.

Another argument against a reptilian identity is that the Type III Multiple Humped Animal's observed vertical flexure is more consistent with mammalian movement. Again, however, the fossil record and examples of extant species do provide precedents and demonstrate vertical flexure among the reptiles. After examining the cervical vertebrae of elasmosaurs, Zammit *et al.* (2008) estimated that these extinct animals had approximately 75-177° flexure in the ventral plane and approximately 87-155° flexure in the dorsal plane. Following his examination of a fossil juvenile pliosaur, Kear (2007) suggested that the steeply angled zygapophyseal articulations on the cervical vertebrae suggest a capacity for vertical movement in the neck. Certainly, modern monitor lizards, turtles, and snakes exhibit recognisable vertical flexure.

Cetacean locomotion is aided by a two-joint system that allows the fluke to be moved independently from the anterior tail. The first joint occurs behind the lumbar region, which is stiff due to the presence of dorsal and ventral ligaments. These ligaments terminate in the anterior tail, giving flexibility to this region. The second region of flexibility is created by large intervertebral spacing and convex articular surfaces (Rommel, 1990). Both of these joints increase the flexibility of the tail along the sagittal plane. Similar two-joint systems, with flexibility provided by different adaptations to the axial skeleton, have been identified in scombroid fishes, such as the swordfish *Xiphias gladius* (Fierstine and Walters, 1968), and in lamnid sharks (Reif and Weishampel, 1986). In these animals, however, flexibility is increased along the horizontal plane.

A significant portion of the vertical flexure argument of the Type III Animal's reptilian classification appears to be associated with the vertical undulations reported in the swimming motion and the horizontally lobed tail. The animal has been observed with the body oriented in vertical loops, or humps, and when the swimming motion has been described, the first hump directly behind the head is reported to remain relatively static (Champagne, 2007). Interestingly, historical accounts of the unidentified animal in Loch Ness describe a surface swimming motion with a changing profile of single to multiple humps, or no undulations visible at all (Costello, 1974). The author (Champagne, 2006) has stabilised a video recording of what appear to be multiple specimens of the Type III Multiple Humped Animal in a possible lekking area in San Francisco Bay, and noted that the multiple humps of fully extended animals did not appear to undulate and maintained their relative positions.

After reviewing the observations by Buffetaut (1983) of the similarity between the vertebrae of modern whales and Jurassic marine crocodilians, LeBlond and Bousfield (1995) suggested that vertical spinal flexure may not be sufficient as a unique, classification characteristic of mammals. Perhaps the animal does not swim with sinuous undulations, but instead remains moderately rigid except for the jointed tail region, with visible humps only providing the appearance of vertical undulation.

Coleman and Huyghe (2003) proposed that small and shallow lakes may be too hot, too polluted, would not provide enough food, do not have sufficient water year round, and can be man-made, obstructed, or both, and are therefore less likely to support "monsters". Such arguments are sound and common as they are applied to an animal with a limited or specialised diet and environment. Certainly, if the undescribed animals were mammalian, their higher metabolic rates would require supportive feeding opportunities and increase their potential visibility, and, by extension, exclude them from consideration in smaller or less optimal bodies of water. Frequently, observations imply any proposed animal would be exclusively piscivorous, although similar, unidentified marine animals have been observed eating birds and small invertebrates (Costello, 1974; LeBlond and Bousfield, 1995; Kirk, 1998; Champagne, 2007). Watson (2013) has discussed the possibility that land sightings of Nessie may be the result of that animal leaving the water to ambush and prey upon deer, but he also listed observations of animals similar in appearance to the Type III Animal eating plants and possibly algae. Others have suggested that proposed, similar animal types may filter particulates from the water column or mud (Mackal, 1976) or maintain a varied diet (Mardis, 2014).

Whereas Mangiacopra (1992) estimated populations of the more commonly reported unidentified lake animals as residents, he and others (Shuker, 1996; Burchfiel, 2012) also allowed for migrant populations. If populations of inland water animals were (or included) transient or migrant members, arguments against acceptable or consistent water quality or conditions, and standing food stocks have less significance on an animals' potential presence when allowing for conditional or seasonal movements. Perhaps, these animals move from marine environments, or between freshwater streams, ponds, or lakes, to winter, birth, and/or exploit seasonal feeding opportunities. In these circumstances, the animals may feed minimally, specifically, or not at all.

Additionally, like the marine Type III Multiple Humped Animal proposal, the sub-terminal to ventral mouth position suggests that the animal feeds primarily at or near the substrate. The Type III Multiple Humped Freshwater Animal may primarily graze on any combination of aquatic plants, algae, benthic invertebrates and fishes, and their substrate-attached or oriented eggs, larvae, and fry. Possibly the animal also comes out of the water to graze on terrestrial plant material or forage for terrestrial organisms.

Historic reports and surveys occasionally imply an amphibious capability to some unidentified freshwater animals (Costello, 1974; Eberhart, 2002; Watson, 2011). Kojo (1991) acknowledged the physical barriers or impediments to presently-unknown freshwater animals' movement, but also left open the possibility that some animals are capable of overland travel and could potentially negotiate obstacles. Shuker (2012) suggested that the undescribed animals observed in Irish loughs and Scottish lochs are temporary residents, likely moving from one body of water to another.

Obviously, for such animals to move between bodies of water their body plan must provide for minimal water levels or terrestrial locomotion. The Type III Multiple Humped Animal appears to have such a capability, based on reported marine and inland observations, having been observed out of the water and in sloughs and shallow water estuaries more than 5 km inland - more frequently than the proposed plesiosaurid body plan.

The Type III Multiple Humped Animal swims and moves with variable, vertical undulations of its tail, possibly providing a minimal profile to any predator or potential observer when travelling overland or through smaller waterways. Reduced (and possibly rayed or jointed) pectoral appendages may allow for negotiating relatively narrow spaces and waterways, as opposed to movement with horizontal undulations, rigid appendages, or both. The tail is distinct and an identifying feature of the type (Fig. 8).

This horizontally-oriented, rayed tail not only provides the propulsion required for open water swimming, as previously proposed, but also may provide an anchor or contact point in terrestrial or shallow water travel. The plated tail may allow the rayed fins to collapse or adjust, and provides a structural support that the animal may push against in order to move through smaller spaces and reduced water levels when swimming is not advantageous or possible. This mode of movement may perhaps be similar to snake concertina locomotion. Possibly supportive, the MacLellan and Grant land sightings of the animal associated with Scotland's Loch Ness displayed "fanned out" rear legs underneath the animal, "seal-wise", and a loping movement similar to a sea-lion (Costello, 1974). As the Type III Multiple Humped Animal has been reported to make a rattling noise or vibration with its plated tail, perhaps the noise is also the source of the purported and disputed echolocations recorded in Lake Champlain, and the "tapping" and "clicking" noises reportedly associated with Nessie in Loch Ness. An extended, semi-retracted, or collapsed tail may also provide any variation in descriptions of the tail during observations.

The serpentine, horse-headed Water Horses of the British Isles reportedly live in oceans, lakes, rivers, and streams, and are observed most frequently in November, leaving a "slimy trail" on land, whereas the Kelpies were believed to cause their home lakes to swell and flood. Additionally, the long-bodied South American minhocão is believed to make water channels with its terrestrial movement, and the "monstrous snake" of the Greek Kafyes marshes was thought to have excavated trenches and left trails in areas where it passed (Vembos, 2014).

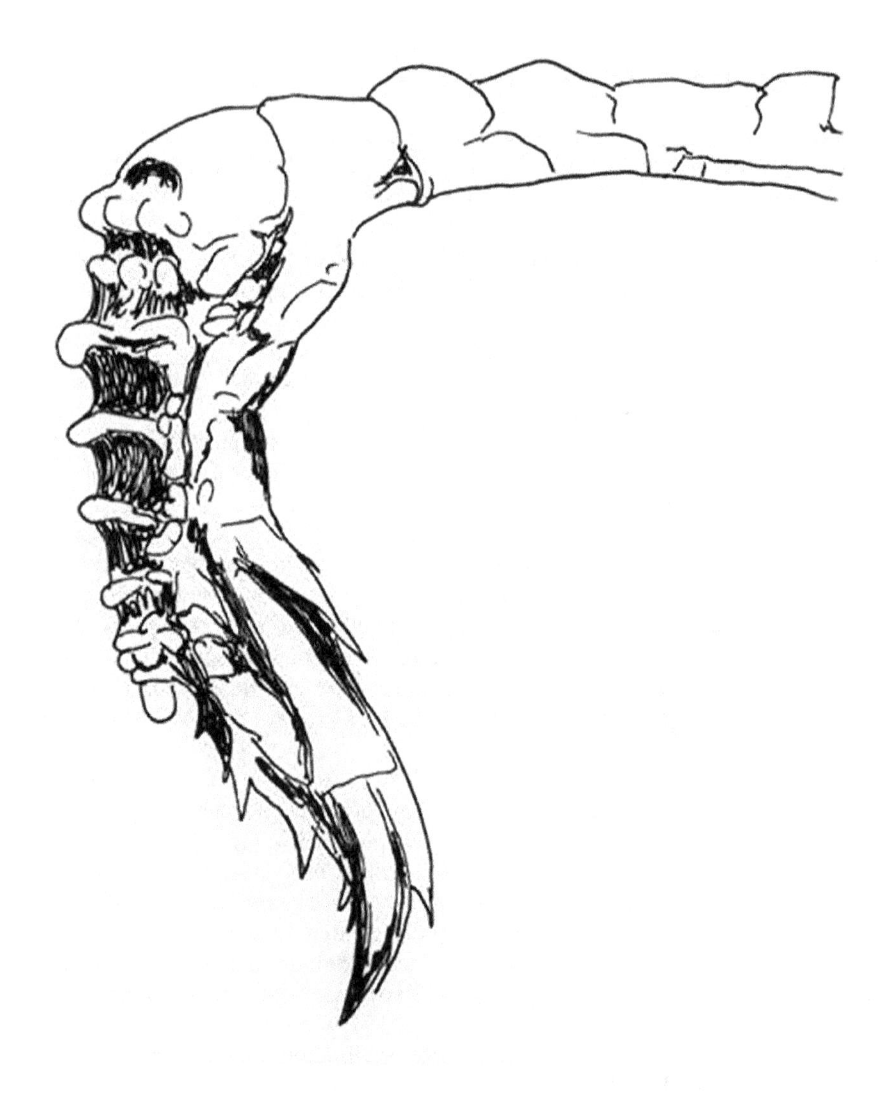

Fig. 8: Type III Multiple Humped Animal tail structure (© Darren Naish)

Similar, unidentified freshwater animals from Siberia, Africa, China, Malayasia, France, Italy, Norway, Sweden, Poland, and Switzerland have also been reported (Eberhart, 2002).

The examination by Watson (2012) of Nessie land sightings noted February as the most frequent month for observations, comprising 25% of all reports where the month is known. Although Watson (2013) advised tnat he has found no correlation between wet weather and land observations, reports have included similar inland animals becoming trapped in culverts and low-volume streams. Perhaps the animal is not the legendary source of flooding and channels, but evidence that the animal is more likely to attempt transiting areas during the rainy season or periods of high water and flooding through existing or seasonal waterways. Such movements may also provide seasonal access to potential mates, and provide for sustainable population distributions.

Other statements against the existence of unidentified lake animals suggest such animals should be observed and reported more frequently (Radford and Nickell, 2006), although Bauer (2002a) argues that even the most frequently "researched" bodies of water have not been surveyed to the thorough extent that critics may suppose. Such critical arguments appear to assume a describable, measurable, or known rate of observational effort, although the author is only familiar with informal assumptions. This unsupported, observational argument would hinge on the number of observers watching and paying direct attention to the entire surface at all times, having the visual acuteness and proximity to recognise unusual surface disturbances or other events, and then to recognise and report or clearly photograph the event. Such a proposal may be based primarily on an animal that regularly disrupts the surface for respiration or some other surface-oriented behaviour.

Certainly, if such animals were cetaceans, one may expect to see and hear exhalations at the surface. Whereas cetaceans may exhale audibly and visibly, pinnipeds and sirenians may surface and breathe unnoticed (Riedman, 1990), with just their nostrils exposed, although pinnipeds also haul out on land for periods of time. If the proposed animals were reptilian, as with the Type III Animal proposal, one might not expect to observe the animal breathing at the surface commonly, as evidenced by the minimal, visible exposure at the surface of modern reptiles like snakes, turtles, and crocodilians.

TYPE III MULTIPLE HUMPED FRESHWATER ANIMAL VARIANTS

The author proposes that the Type III Multiple Humped Freshwater Animal could be differentiated into at least three variants, one of which may also include the marine form (Fig. 9). A "resident" form may be possible in environments with relatively stable resources and levels of competition and predation. These environments may also provide for and require a comparatively larger, somatic body size, generalised feeding strategies, slower metabolism, and greater temperature tolerances. Any resultant low genetic diversity could be addressed if these resident animals had breeding roles, with only a minimal number of individuals actually mating, or with another variant. Perhaps this form has become behaviourally differentiated in response to human activity or a specific environment, as noted between morphologically similar types with dissimilar temporal behaviours.

VARIANT	POPULATION	RESOURCES	COMPETITION	PREDATION	MISCELLANEOUS
Resident	Stable	Stable	Stable	Stable	Medium sized, overland movement not required
Transient	Variable	Variable	Variable	Variable	SM-MD body size/type allows for overland movement
Migrant	Variable	Predictable	Variable	Stable	MD-LG size/type allows for some overland movement
Estuarine	Variable	Stable	Variable	Variable	Some restriction in body size and type
Isolated Lake	Stable	Stable	Stable	Stable	Body size/type less restricted
Unrestricted River/Lake	Variable	Variable	Variable	Variable	Restriction in body size and type

Fig. 9: Comparison of Type III Freshwater Animal Variants

A second variant, a "migrant" form, would naturally have varied distribution, dependent on temporal and spatial conditions, as the animal travels to exploit seasonal or cyclical conditions, limited or specific feeding resources, or areas with increased mating opportunities. Under such circumstances, intraspecific competition could be expected to be variable, and at times potentially high. Levels of predation could also be high, depending on the significance of the event, resource, or point of migration; although a diluting effect may also be provided by the increased number of conspecifics in a specific area. A migrant variant could have a reduced body size that would allow for varied movements, both terrestrial and aquatic, in order to realise migratory objectives. However, some natural or man-made environmental features may limit unrestricted aquatic movements and habitats. Perhaps the Type III Multiple Humped Freshwater Animal migrant variant is more closely related to the marine type.

Possible lekking areas are regularly reported for morphologically similar animals off the northern west coast of North America and the British Isles, with similarly described animals reported more than 5 km inland, when access from a marine environment has not been restricted (Champagne, 2007). As this variant may move from marine to brackish and freshwater environments, additional arguments against this variant's reptilian assignment have included the observation that mammals are far more euryhaline than reptiles. Again, the fossil record and an examination of extant reptile species provide precedents in response to this argument.

Multiple species of extant crocodilian (estuarine crocodile *Crocodylus porosus,* Nile crocodile *Crocodylus niloticus,* American crocodile *Crocodylus acutus,* Orinoco crocodile *Crocodylus*

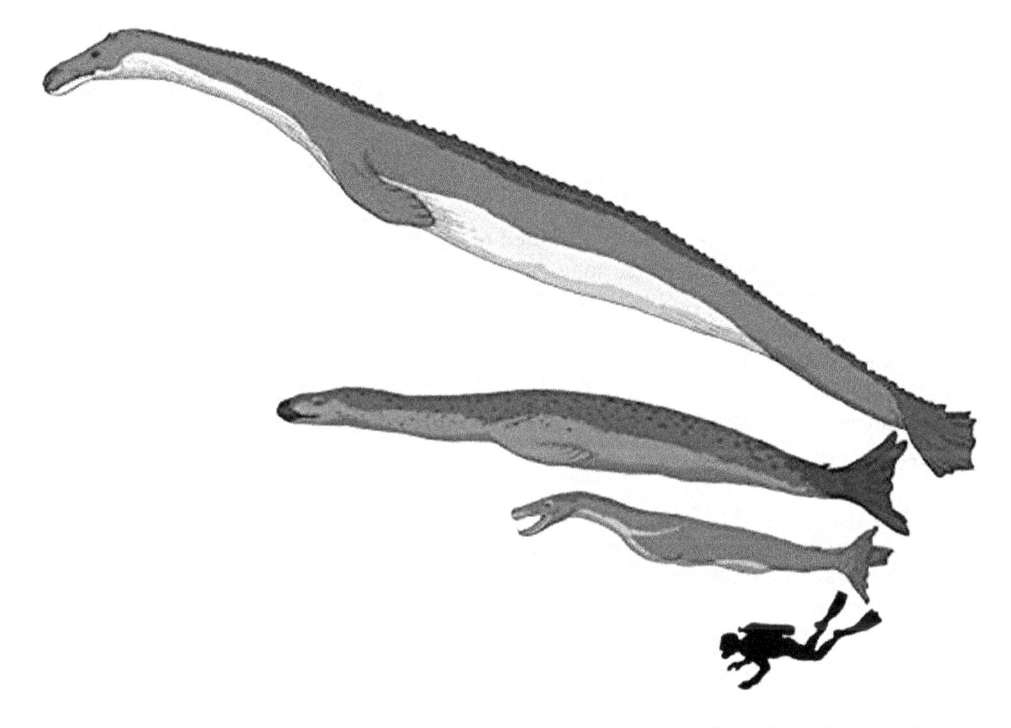

Fig. 10: Three possible still-undiscovered pinnipeds (© Michael Woodley *et al.*)

intermedius, Indian mugger crocodile *Crocodylus palustris*), the diamondback terrapin *Malaclemys terrapin,* Asiatic river snakes of the taxonomic family Hamalopsidae, monitor lizards (Varanidae), file snakes (Acrochordidae), and even skinks (the mangrove skink *Emoia atrocostata*) are known to inhabit and move between waters of varied or no salinity (Bellairs, 1970). The Eocene fossil record contains multiple, probably aquatic, snake species simultaneously inhabiting marine, brackish, and freshwater environments (Rage *et al.*, 2003). Following an examination of fossils from Mesozoic deposits in the Canadian Arctic archipelago, Vavrek *et al.* (2014) suggested not only that high latitude polycotylid plesiosaurs inhabited near-shore waters, but also that they periodically forayed into freshwater environments. Holmes *et al.* (1999) proposed a new species of plioplatecarpine mosasaur, with supporting deposits, and was suggested as exploiting both estuarine and freshwater environments. Kear (2007) also discussed pliosaur fossils suggesting their simultaneous presence in the marine, brackish, and freshwater environments of the Australian Cretaceous epicontinental seaway. While discussing the rapid evolution and adaptive radiation of mosasaurs, Everhart (2005) suggested that mosasaurs may not only have contributed to the extinction of several shark species, but also provided competition for crocodiles in estuarine and freshwater environments.

A third Type III Multiple Humped Freshwater Animal variant could be described as "transient". Such a variation could be expected in areas with variable resources, competition, and predation, and a required generalised approach to habitat and diet. A relatively smaller body size, indicative of unpredictable feeding opportunities and potential greater energy expenditures, would also allow the animal to negotiate and mitigate environmental obstacles and conditions, and provide a variable, although possibly philopatric, distribution.

Interestingly, Woodley *et al.* (2009) attempted to estimate the number of pinniped species remaining for formal description. The authors referenced three types of sea serpents proposed or referenced in cryptozoological literature - the long necked sea serpent, merhorse, and tizheruk, and morphologically compared them to the animal known as *Cadborosaurus willsi*, also proposed by the author as a Type III Multiple Humped Animal (Champagne, 2007). Each of the authors' pinniped proposals recognised a size class, which have an appearance and geographic range comparing favorably to the Type III Multiple Humped Freshwater Animal and variant proposal (Fig. 10).

Some natal sympatry may be present between Type III freshwater variants and the marine type, and observed individuals may be representative of an overlapping age class or distributed through an uncommon environmental event. However, based on the size distributions of the general descriptions, the author estimates the size ranges and likely environments for the resident variant as 6-10 m TBL, and more likely to be reported in isolated, relatively larger bodies of water. The transient variant is reported most frequently in a size range of 3-5 m TBL, and most commonly associated with unrestricted or minimally obstructed rivers, lakes, and inland freshwater marshes. The migrant variant may be typically assigned to a size class in excess of 10 m TBL, and likely to be reported in relatively larger, unobstructed rivers and lakes.

Reproduction may occur between overlapping variants and/or the marine type, or speciation may have occurred in one or more of the variants. Perhaps similarly, Pittman and Ensor (2003) suggested that the low genetic diversity in multiple forms of Antarctic killer whales (*Orcinus orca*) were the result of rapid and recent divergence from the phenotype- possibly in as little as thirteen generations.

CONCLUSION

Variations of the Type III Multiple Humped Freshwater Animal proposal appear to correspond with the different environments associated with unidentified freshwater animal (or "lake monster") reports. Each variant with its restrictive or supportive body size, behaviour, and morphology appears consistent with reports, descriptions, and the observations of other researchers, local legend, or both.

The author estimated that the marine Type III Multiple Humped Animal attains the largest size, and would likely be found only in inland waterways with direct access to the ocean. The marine variant does, however, appear to maintain some ability to move onto land short distances from the water.

Attaining a TBL size in excess of 10 m, the "migrant" Type III Freshwater Animal variant is relatively smaller than the marine type, but has a body size allowing for more diverse movements, both terrestrial and aquatic, in order to exploit migratory resources, and it is more likely to be a candidate for reports from larger unrestricted or minimally restricted bays, lakes and rivers.

The "resident" Type III Freshwater Animal variant appears intermediate in size, possibly attaining a TBL of 6-10 m. The author proposes this variant to be the probable candidate for unidentified animals in larger, isolated lakes and other freshwater bodies.

Within the 3-5-m TBL size class, the relatively smaller "transient" Type III Freshwater Animal variant could be expected in inland freshwater systems with variable or low flows, or with natural, seasonal, or man-made obstacles. This type variant would most likely be reported in very shallow waterways, and observed with terrestrial movements.

The Type III Multiple Humped Freshwater Animal (and possible variants) proposal compares favorably with the reported, historical, and artistic morphology of both prominent and lesser known undescribed freshwater animals. As the Type III Animal displays activity and agreeable morphology that encompasses most reported behaviours, composite serpentine and plesiosaurid representations depicting long necks, equine-like or cameloid heads, unnatural and inexplicable illustrations of humps, and a fluked tail become unnecessary to account for the varied descriptions, if one considers the Type III Animal and variant proposal.

It was not the author's specific intent to prove or disprove the existence of a particular "lake monster" or type proposal. Rather, by comparing and correlating various candidate proposals, viable population estimates, access, and available resources with the proposed Type III Multiple Humped Animal and variants, a possible natural history and viability can be examined and falsified, maximizing infrequent occasions for observations and limited resources to collect and analyse additional data for eventual specimen procurement, tissue collection, or both..

ACKNOWLEDGMENTS
I am appreciative of Gary Mangiacopra, who has always been helpful and extremely generous with his information, insights, and resources, as well as Roland Watson. Few researchers have such a command of the relevant and supportive literature and existing database as Scott Mardis. I am grateful to him for his input and generosity. Tim Morris was most gracious in providing his talents in illustrating the proposals in this work. Dr Karl Shuker's encouragement and patience was also greatly appreciated, as were the anonymous reviewers.

REFERENCES

- Andrews, R., (1935). *Roy Chapman Andrews and the Double-Finned Whale.* G.P. Putman's Sons (New York).
- Bartholomew, R. (2012). *The Untold Story of Champ: A Social History of America's Loch Ness Monster.* State University of New York Press (Albany).

- Bauer, H., (2002a). Common knowledge about the Loch Ness Monster: Television, videos, and films. *Journal of Scientific Exploration*, 16(1): 455-477.
- Bauer, H., (2002b). The case for the Loch Ness "Monster": The scientific evidence. *Journal of Scientific Exploration*, 16(2): 225-246.
- Bellairs, A. (1970). *The Life of Reptiles.* Universe Books (New York).
- Bernard, A., *et al.* (2010). Regulation of body temperature by some Mesozoic marine reptiles. *Science*, 328 (No. 5984; 11 June): 1379-1382.
- Buffetaut, E. (1983). Vertical flexure in Jurassic and Cretaceous marine crocodilians and its relevance to modern 'sea serpent' reports. *Cryptozoology*, 2: 85-89.
- Burchfiel, K. (n.d.). Rogue Nessie. Strangemag.com, http://www.strangemag.com/roguenessie.html. Retrieved 07-17-12.
- Brisbin, I.L., *et al.* (1982). Body temperature and behaviour of American alligators during cold winter weather. *American Midland Naturalist,* 107: 209-218.
- Champagne, B. (2001). A preliminary evaluation of a study of the morphology, behaviour, autoecology, and habitat of large, unidentified marine animals, based on recorded field observations. *Crypto-Dracontology Journal*: 93-112.
- Champagne, B. (2006). A preliminary examination of a stabilized video recording of large, unidentified animals in San Francisco bay recorded on January 26, 2004. Cryptomundo, http://cryptomundo.com/wp-content/uploads/framestabilization.pdf
- Champagne, B. (2007). A classification system for large, unidentified marine animals based on the examinations of reported observations. *In:* Heinselman, C. (Ed.), *Elementum Bestia*. Lulu (Raleigh): 144-172.
- Colbert, E.H., *et al.* (1946). Temperature tolerances in the American alligator and their bearing on the habits, evolution and extinction of the dinosaurs. *Bulletin of the American Museum of Natural History,* 86: 327-374.
- Coleman, L. (2001). Lake monsters. *Crypto-Dracontology Journal.*
- Coleman, L. (2009). On Site: Lake Champlain 2009. Cryptomundo, http://www.cryptomundo.com/cryptozoo-news/champ-loren09/ 23 June. Retrieved 19 June 2012.
- Coleman, L. and Huyghe, P. (2003). *Field Guide to Lake Monsters, Sea Serpents, and Other Mystery Denizens of the Deep.* Tarcher/Penguin (New York).
- Costello, P. (1974). *In Search of Lake Monsters.* Garnstone Press (London).
- Couper, A., (1983). *The Times Atlas of the Oceans.* Van Nostrand Reinhold Company, Inc. (New York).
- Eberhart, G. (2002). *Mysterious Creatures: A Guide to Cryptozoology.* ABC-CLIO, Inc. (Santa Barbara).
- Elizabeth, K. (2014). Personal communications. 25 May and 4 August.
- Elizabeth, K. and Hall. D. (2014). *Water Horse of Lake Champlain II.* CreateSpace Independent Publishing Platform.
- Ernst, C. and Lovich, J. (2009, 2011). *Turtles of the United States and Canada* (two editions). Johns Hopkins University Press (Charles Village, Baltimore).
- Everhart, M.J. (2005). Rapid evolution, diversification and distribution of mosasaurs (Reptilia; Squamata) prior to the K-T boundary. *11th Annual Symposium in Paleontology and Geology*: 16-27.

- Fierstine, H. and Walters, V. (1968). Studies in locomotion and anatomy of scombrid fishes. *Memoirs of the Southern California Academy of Sciences*, 6: 1–31.
- Frizzell, M. (2001). The Chesapeake Bay serpent. *Crypto-Dracontology Journal.*
- Hall, D. (2000). *Champ Quest 2000 the Ultimate Search: Field Guide and Almanac to Lake Champlain.* Essence of Vermont (Jericho).
- Harrison, P. (2001). *Sea Serpents and Lake Monsters of the British Isles.* Robert Hale (London).
- Heuvelmans, B. (1968). *In the Wake of the Sea-Serpents.* Hill and Wang (New York).
- Holmes, R., *et al.* (1999). A new specimen of *Plioplatecarpus* (Mosasauridae) from the lower Maastrichtian of Alberta; comments on allometry, functional morphology, and paleoecology. *Canadian Journal of Earth Sciences*, 36(3): 363-369, 10.1139/e98-112.
- Ingmanson, D. and Wallace, W., (1985). *Oceanography: An Introduction.* Wadsworth Publishing Co. (Belmont).
- Kear, B. (2005). Marine reptiles from the Lower Cretaceous (Aptian) deposits of White Cliffs, southeastern Australia; implications of a high latitude, cold water assemblage. *Cretaceous Research*, 26: 769-782.
- Kear, B. (2007). A juvenile pliosaurid plesiosaur (Reptilia: Sauropterygia) from the Lower Cretaceous of southern Australia. *Journal of Paleontology*, 81(1): 154-162.
- Kirk, J. (1998). *In the Domain of the Lake Monsters.* Key Porter Books Ltd (Toronto).
- Kojo, Y., (1991). Some ecological notes on reported large, unknown animals in Lake Champlain. *Cryptozoology*, 10: 42-54.
- Kojo, Y. (1992). Distributional patterns of cryptid eyewitness reports from Lake Champlain, Loch Ness, and Okanagan Lake. *Cryptozoology*, 11: 83-89.
- LeBlond, P. and Bousfield, E., (1995). *Cadborosaurus: Survivor from the Deep.* Horsdal and Schubart (Victoria, B.C.).
- Mackal, R.P. (1976). *The Monsters of Loch Ness.* The Swallow Press (Chicago).
- Mackal, R. (1980). *Searching for Hidden Animals: An Inquiry into Zoological Mysteries.* Doubleday and Company (Garden City).
- Mangiacopra, G. (1992). *Theoretical Population Estimates of the Large Aquatic Animals in Selected Freshwater Lakes of North America.* Southern Connecticut State University, New Haven, Connecticut.
- Mardis, S. (2014). Personal communications. 12 February, 25 March, 22 May.
- Mardis, S. (2015). Personal communication. 31 March.
- National Geographic (1990). *National Geographic Atlas of the World.* National Geographic Society (Washington D.C.).
- Nikolsky, G.V. (1978). *The Ecology of Fishes.* TFH Publications (New Jersey).
- O'Neill, J. (1999). *The Great New England Sea Serpent: An Account of Unknown Creatures Sighted by Many Respectable Persons Between 1638 and the Present Day.* Down East Books (Camden).
- Oudemans, A. (1892). *The Great Sea Serpent.* Luzac and Co (London).
- Paxton, C. (2010). Personal communication. 17 July.
- Pittman, R. and Ensor, P. (2003). Three forms of killer whales (*Orcinus orca*) in Antarctic waters. *Journal of Cetacean Research and Management*, 5(2): 131–139.

- Radford, B. and Nickell, J. (2006). *Lake Monster Mysteries: Investigating the World's Most Elusive Creatures.* The University Press of Kentucky (Lexington).
- Rafinesque-Schmaltz, C. (1819). Dissertation on water-snakes, sea-snakes and sea-serpents. *Philosophical Magazine*, 54 (November): 361-367.
- Rage J.-C., *et al.* (2003). Early Eocene snakes from Kutch, Western India, with a review of the Palaeophiidae. *Geodiversitas*, 25(4): 695-716.
- Reif, W. and Weishampel, D. (1986). Anatomy and mechanics of the lunate tail in lamnid sharks. *Zoologische Jahrbücher. Abteilung für Anatomie und Ontogenie der Tiere...,* 114: 221-234.
- Riedman, M. (1990). *The Pinnipeds: Seals, Sea Lions, and Walruses.* University of California Press (Berkeley).
- Rommel, S. (1990). Osteology of the bottlenose dolphin. *In:* Leatherwood, S. and Reeves, R.R. (Eds), *The Bottlenose Dolphin.* Academic Press (London): 29-49.
- Scheider, W. and Wallis, P. (1973*).* An alternate method of calculating the population density of monsters in Loch Ness. *Limnology and Oceanography*, 18: 343.
- Sheldon, R. and Kerr, S. (1972). The population density of monsters in Loch Ness. *Limnology and Oceanography*, 17: 796-798.
- Shuker, K. (1996). *The Unexplained.* JG Press (North Dighton).
- Shuker, K. (2012). Untitled post. Journal of Cryptozoology Facebook Group, https://www.facebook.com/groups/261237557287228/ Posted 5 June.
- Smith, T. and Horonowitsch, G. (1987). Harbour seals in the Lacs des Loups Marins and Eastern Hudson Bay Drainage. *Canadian Technical Report of Fisheries and Aquatic Sciences*, No. 1536: 17 pp.
- Storer, T.I., *et al.* (1979). *General Zoology*. McGraw-Hill (New York).
- Vavrek, M., *et al.* (2014). Arctic plesiosaurs from the Lower Cretaceous of Melville Island, Nunavut, Canada. *Cretaceous Research*, 50: 273e281.
- Vembos, T. (2014). Haunted snakes and lake monsters of Greece. Vembos, http://www.vembos.gr/hauntedsnakes.htm (n.d.). Retrieved 16 October 2014.
- Von Muggenthaler, E., *et al.* (2010). Echolocation in a freshwater lake. *Journal of the Acoustical Society of America*, 127 (23 March): http://scitation.aip.org/content/asa/journal/jasa/127/3/10.1121/1.3384449
- Watson, R. (2011). *The Water Horses of Loch Ness.* CreateSpace (London).
- Watson, R. (2012). Nessie on land: the overview. Loch Ness Monster, http://lochnessmystery.blogspot.co.uk/2012/05/nessie-on
- land-overview.html 16 May. Retrieved on 18 May 2012.
- Watson, R. (2013). Personal communications. 18 and 28 November.
- Woodley, M., *et al.* (2009). How many extant pinniped species remain to be described? *Historical Biology*, 20(4): 225-235.
- Zammit, M., *et al.* (2008). Elasmosaur (Reptilia: Sauropterygia) neck flexibility: implications for feeding strategies. *Comparative Biochemistry and Physiology, Part A,* 150: 124-130.

APPENDIX

The following figures summarise the data amassed by the author as discussed above.

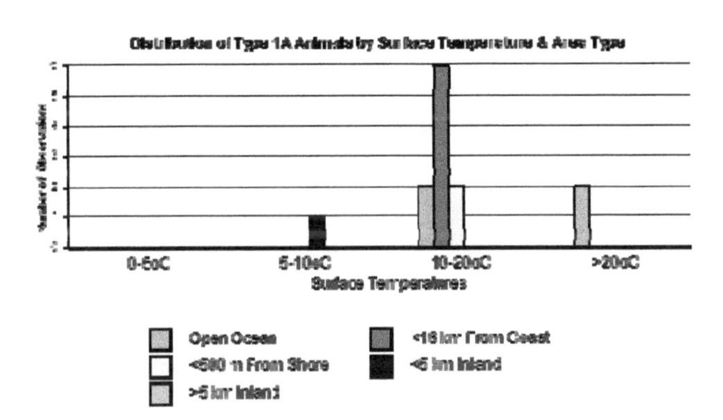

Fig. 11: The preliminary examination of the distribution of Type 1A Animals (Long Neck variation) by the prevailing surface temperatures and designated area type. Each bar contains the total of the observations reported (N=13) (*Crypto-Dracontology Journal*).

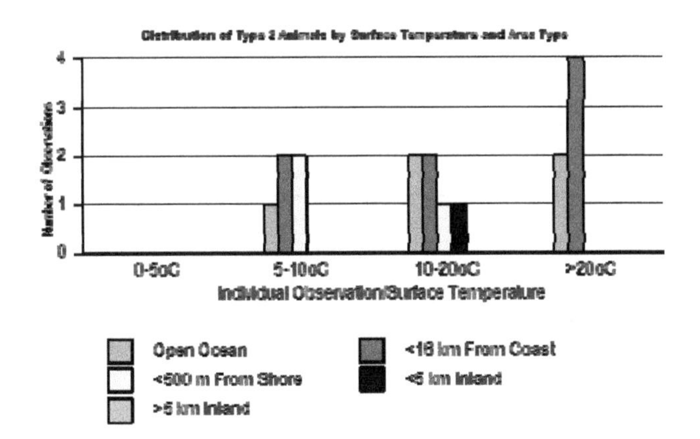

Fig. 12: The preliminary examination of the distribution of Type 2 Animals (Eel-Like) is illustrated by the prevailing surface temperature and area type. Each bar contains the total of observations reported (N=17) (*Crypto-Dracontology Journal*).

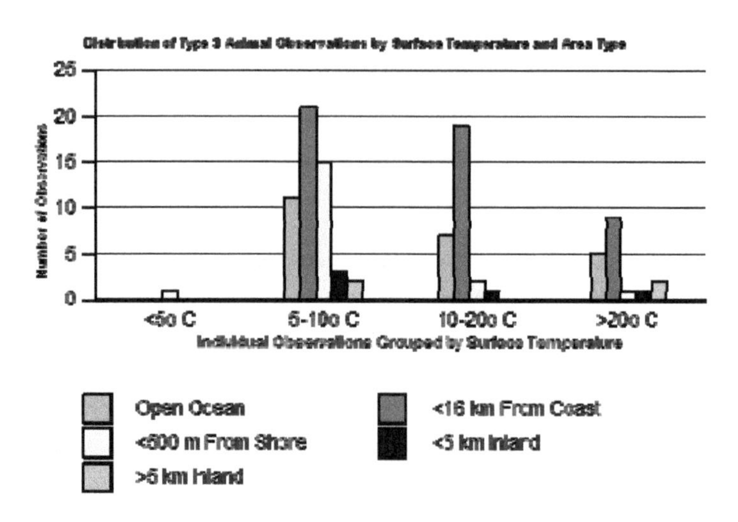

Fig. 13: The preliminary examination of the distribution of Type 3 Animal (Multiple-Humped) observations by the surface temperature and the area type in which the observation occurred. Each bar represents the total of observations (N=99), for that temperature range and area type (*Crypto-Dracontology Journal*).

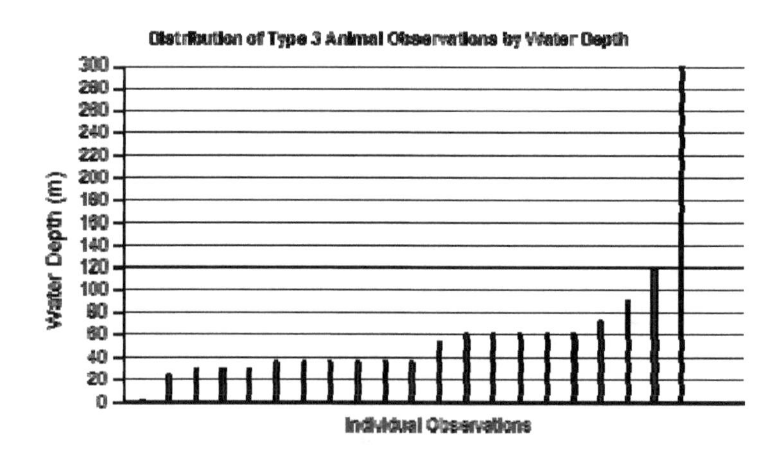

Fig. 14: The preliminary examination of the distribution of Type 3 Animal (Multiple-Humped) observations by the estimated depth in the area the observations occurred. Each bar represents a single observation (N=21). Two open ocean observations are not illustrated. The depths for those observations were 810 m and 4500 m. (*Crypto-Dracontology Journal*).

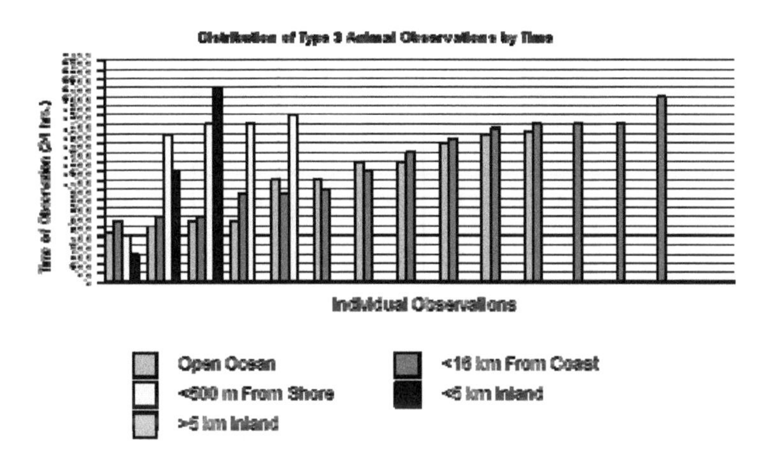

Fig. 15: The preliminary examination of the distribution of Type 3 Animals (Multiple-Humped) by Time (24-hour clock) and the Area Type of the reported observation. Each bar represents a separate observation (N=34) (*Crypto-Dracontology Journal*).

OVERLEAF:

Table 1: The preliminary examination of the comparison of *C. willsi*, the Great New England Sea Serpent (GNESS), an unidentified animal observed in proximity to the United Kingdom, countries of Scandinavia, and Indochina. and of the Pacific Northwest, with the Type 3 Animal (Multiple Humped) (*Crypto-Dracontology Journal*).

CHARACTERISTICS	TYPE 3	*C.willsi*	GNESS	U.K.	SCANDINAVIA	INDOCHINA
Range (latitude)	30-60o	35-60o	30-45o	49-58o	56-71o	24-46o
Season(s) Observed	ALL	ALL	Summer	Summer, Spring	Summer	Summer, Winter
Area Type	1,2,3,4, 5, land	1,2,3,4, 5, land	1,2,3,4, 5, land	1,2,3,4	2, 3	2,3,4
Eccritic Water Temp. (oC)	5-20o	5-12o	17.5o	7.5-15o	10-17.5o	10o
Swimming Method	Humped, w/ vertical undulations	Humped, w/ vertical undulations	Humped, w/ vertical undulations	Humped, w/ vertical undulations	Humped, w/ vertical undulations	Humped, w/ vertical undulations
Total Body Length	3-60m	5-15m	3-40m	10-80m	20-80m	20-25m
Head Description	Cameloid	Cameloid	Snake-like	Horse, alligator, snake, flat	Horse, alligator, long, blunt, pointed	Snake, seal
Body Description	Serpentine	Serpentine	Serpentine	Serpentine	Serpentine	Serpentine
Tail Description	Horizontal, planed	Horizontal, planed	Horizontal, forked	Forked, opaque	Unknown, but emits clapping noise	Unknown
Fin Description	Pectoral only	Pectoral only	2 pair?	"Flappers"	Unknown	Unknown
Dorsal Structures	Serrations, mane	Serrations, mane	Serrations	Serrations	Serrations, mane	Serrations
Type 3 Agreement	N/A	100%	82%	91%	91%	73%

INSTRUCTIONS TO CONTRIBUTORS

These fall into two categories: important issues to consider when preparing a cryptozoological paper; and the style of presentation required for submissions to the journal.

(i) Important Issues to Consider When Preparing a Cryptozoological Paper

The *Journal of Cryptozoology* aims to publish papers of equal rigour to existing mainstream science journals. Cryptozoology is a controversial topic. Consequently, the following guidelines suggest important issues to consider when preparing a cryptozoological paper, in order to pre-empt common criticisms that might be made by reviewers.

Premises and Background

1 The scientific literature works from the existing consensus. By definition (i.e. Heuvelmans, 1988), there is no consensus on whether the putative animals behind cryptozoological reports exist. Therefore papers submitted to the *Journal of Cryptozoology* should *not* start from a premise that a specific animal species does exist (except to make predictions about the evidence for it that might be found in future or to develop methodology). Papers can of course present evidence for the existence of unknown animals, and argue against the existing consensus.

2 Papers should work from the existing scientific consensus with regard to other arguments as well. This does not mean non-consensual assumptions cannot be made but that if such assumptions are made, they should be justified by reference to evidence. For example, if an author believes that certain bipedal tracks are made by a putatively still-extant prehistoric animal that palaeontologists assume is quadrupedal (and extinct), then some evidence for the presumed animal's bipedality, *independent* of the tracks, should be given.

3 Professions of personal belief in particular cryptids are not part of a scientific paper, but personal experience of seeing putatively unknown animals may be suitable for publication.

4 It is particularly important that authors should *not* presume the zoological identity of a particular cryptid (see also Point 17 below), nor should language be employed that implies the identity is unequivocally known. Papers can of course argue for a particular zoological explanation for a certain set of reports, given particular evidence and test hypotheses concerning particular identities.

5 Authors should be aware of and cite the relevant scholarly literature on the topic in question. The use of *Google Scholar* or similar search engines allows basic searching of the scholarly literature by people outside of research institutions. References used should be scholarly where possible, and definitely from print media or an online equivalent. Ephemeral references should not be used except as an absolute last resort. Non-scientific blogs and similar articles are generally unacceptable as sources in formal papers, as is *Wikipedia.*

6 Authors should be aware that words can have a technical meaning within certain fields that is different from that in general use. For example, the terms 'accuracy' and 'precision' have specific technical meanings in statistics, although the words are often considered synonymous by the general public. Likewise, 'bug' as a strict zoological term is restricted to one particular taxonomic order of insects, as opposed to its more general use in North America for almost any terrestrial arthropod.

Using Eyewitness Testimony

Unlike conventional zoology, cryptozoology often uses eyewitness testimony as a potential source of evidence. Authors should be aware of the numerous biases associated with eyewitness testimony in general (e.g. Loftus, 1996) and in a specifically cryptozoological framework (e.g. Arment, 2004; Paxton, 2009).

7 Given witnesses cannot, by definition, know what it is that they are reporting, they cannot necessarily gauge their own accuracy or precision.

8 Authors should not necessarily assume eyewitnesses are correct in their identification of body parts and/or the taxonomic affinities of what they are reporting. In addition, there may also be ambiguity in the eyewitness reports. For example, is a reported 'mane' a lion-like mane or a horse-like mane or something different from both of these?

9 Just because reports come from a given area does not mean they have the same origin.

10 Just because reports are hypothesised to have the zoological source (i.e. cryptid) does not mean they actually have the same source. For instance, a given report of a bigfoot could be of an unknown animal, of a man in a gorilla suit, or of a variety of known animal. The origins of reports superficially of the same type need not be consistent in space.

11 Catalogued eyewitness reports are an outcome of a long process that includes *acquisition* (i.e. the perception of the original event), *retention* (i.e. memory), *retrieval* (i.e. recollection), transmission, and recording (Loftus, 1996; Paxton, 2009). At each and every stage, bi-

ases may creep in, which means that accessible reports may represent a very inaccurate and imprecise sample of what was actually seen.

12 From Point 11, it is very possible that collected cryptozoological reports suffer from sampling biases that affect the conclusions drawn from them. For example: if aquatic cryptids are reported most in calm weather conditions (e.g. Mackal, 1976), does this reflect the biology of an unknown animal or is it due to increased detectability under those conditions? (See Buckland *et al.*, 2001 for a discussion of detectability in an animal survey context.)

13 The most frequent source of bias in cryptozoological reports is that they record presence only. If conclusions want to be drawn from reported occurrences or report frequencies, then this can only really be done if search effort (e.g. man-hours searching, area searched, effort in collecting reports, etc) is known.

14 It follows from Point 13 that raw quantification of reports without taking account of search effort *cannot* be used to infer putative habitat preferences of cryptids, etc.

15 Trends in features of reports, if real, may reflect cultural changes of biases in the reporting process rather than changes in the biology of the putative animals, if any.

16 Wherever possible, the original wording of reportees should be used. Authors should avoid interpolating meaning to witness statements unless absolutely necessary.

17 Cryptids are hypothetical constructs of what the putative source of eyewitness reports might be. The raw data of eyewitness-based cryptozoology is reports. Data and hypotheses should not be confused. It is therefore inappropriate to state "bigfoot occurs in the forested areas of British Columbia", when what is incontestably true is "bigfoot reports come from the forested regions of British Columbia".

Reaching Conclusions

All conclusions in the *Journal of Cryptozoology* should be evidence-based - where the premises, data, and chain of reasoning leading to the conclusion are clearly stated. Every point made *must* be justified either through evidence and argument supplied within the paper or by reference to existing published literature. Care should be made in distinguishing between possible explanations of things, of which there are an infinite number, and probable explanations of things, of which there are few and which are of rather more interest to the readership. Probable rival hypotheses for explanations of certain phenomena should be explicitly stated and the reasons for rejecting/accepting a particular hypothesis over others explicitly given.

See also Arment (2004) and Paxton (2011) for more commentary on the methodology of cryptozoology.

References

- Arment, C. (2004). *Cryptozoology, Science and Speculation.* Coachwhip (Landisville).
- Buckland, S.T., *et al.* (2001). *An Introduction to Distance Sampling.* Oxford University Press (Oxford).
- Heuvelmans, B. (1988.) The sources and methods of cryptozoological research. *Cryptozoology*, 7: 1-21.
- Loftus, E.F. (1996). *Eyewitness Testimony.* Harvard University Press (Cambridge [MA]).
- Mackal, R.P. (1976). *The Monsters of Loch Ness.* Futura (London).
- Paxton, C.G.M. (2009). The plural of "anecdote" can be "data": statistical analysis of viewing distances in reports of unidentified giant marine animals 1758–2000. *Journal of Zoology*, 279: 381–387.

Paxton, C.G.M. (2011). Putting the "ology" into cryptozoology. *Biofortean Notes*, 1: 7-20.

(ii) The Style of Presentation Required for Submissions to the Journal

All submissions must be original manuscripts not previously published elsewhere or submitted elsewhere simultaneously with submission to this journal. All submissions will be sent to two members of the journal's peer review panel for their opinions concerning content, clarity, and relevance to cryptozoology. Their comments will then be studied by the editor whose decision is final concerning whether or not the manuscript is published, subject if necessary to amendments by the author(s) if suggested by the reviewers. The copyright of all published papers belongs to this journal.

All manuscripts submitted should be one of the following three types of paper:

Discussion/Review article:

Its subject should be a discussion or literature review of a given cryptozoological subject, and should not include original, unpublished research. It can be of 1000-4000 words in length, and can also include clearly labelled and numbered b/w photographs, artwork, tables, or maps, provided that the copyright of these falls into one of the following three categories:

(1) owned by the author(s);

(2) has been granted to them in writing by their copyright owner(s) - a copy of such permission will need to be submitted with the manuscript and artwork;

(3) expired, i.e. in the public domain.

The article should be preceded by a 200-word abstract, and should be divided into relevant subtitled sections. A reference list can be included at the end of the article; if so, this and the accompanying in-text citation style should correspond with the preferred version outlined below.

Research article:

Its subject should be original research (but not fieldwork) conducted by the author(s). It should be of comparable length to or shorter than discussion/review articles, but with a minimum count of 1000 words. It can also include clearly labelled and numbered b/w photographs, artwork, tables, or maps, provided that the copyright of these falls into one of the three above-listed categories. The article should be preceded by a 100-200 word abstract, and its main text should be split into four sections – Introduction, Materials and Methods (or Description where more appropriate), Results/Interpretation, Discussion/Conclusions. A reference list can be included at the end of the article; if so, this and the accompanying in-text citation style should correspond with the preferred version outlined below.

Field report:

Its subject should be fieldwork conducted by the author(s). It should be of 1000-2500 words in length. It can also include clearly labelled and numbered b/w photographs, artwork, tables, or maps, provided that the copyright of these falls into one of the three above-listed categories. The article should be preceded by a 200-word abstract, and its main text should be split into four sections – Introduction, Description (in which the fieldwork undertaken is described), Results, Discussion (which should also include details of any future plans). A reference list can be included at the end of the article; if so, this and the accompanying in-text citation style should correspond with the preferred version outlined below.

Style of reference citation required:

All in-text citations should be: author(s) surnames, comma, year of publication, all in parentheses. If the cited reference has more than two co-authors, give only the first surname followed by *et al*. Examples: (Jones, 1987), or (Jones and Jones, 1987), or (Jones *et al*., 1987).

For books, the style required for the reference list should be: Author surname followed by given names as initials, then followed by the year of publication in parentheses, and a full stop/period. The title of the book should be italicised, with its principal words beginning with a capital letter, and should end with a full stop/period. The publisher's name should then be given, with the town or city of publication included in parentheses. If the book is co-authored by two authors, their names should be separated by 'and'; if co-authored by more than three, only the first author's name should be given, followed by a comma and then '*et al.*' (in italics). Here are some hypothetical examples:

Smith, J.C. (1987). *The History of Cryptozoology.* Jones and Son (London).

Smith, J.C. and Jones, J.A. (1987). *The History of Cryptozoology.* Jones and Son (London).

Smith, J.C., *et al.* (1987). *The History of Cryptozoology.* Jones and Son (London).

For journal articles, the style required for the reference list should be: Author surname followed by given names as initials, then followed by the year of publication in parentheses, and a full stop/period. The title of the article should not be italicised, and should not be capitalised (other than for the first word or proper nouns). The title of the journal should be given in full, not abbreviated, with its principal words beginning with a capital letter, it should be italicised, and should end with a comma. Volume numbers should be given as figures, issue numbers also as figures (preceded by no.) but included in parentheses following the volume number (together with date of issue if relevant, and separated from issue number by a semi-colon), followed by a colon, and then the page numbers, given in full. If the article is in a newspaper, the town or city of publication in parentheses should follow the newspaper's title, and instead of volume numbers, the full date of publication will suffice, followed by the page number(s) if known. Here are some hypothetical examples:

Smith, J.C. (1987). Investigation of an unidentified lizard carcase discovered in Senegal. *Journal of Lizard Studies*, 33 (no. 2; September): 52-59.

Smith, J.C. (1987). Mystery cat on the loose in Wales. *Daily Exclusive* (London), 4 February: 23.

For online sources, if an author name is given, it should be presented in the same style as for books and articles, followed by the year and title of the source, which should adhere to the style format given above for a hard-copy journal article, followed by the website's title, then the complete URL, date of posting if given, and the date upon which it was accessed by the paper's author(s). Here is an example:

Shuker, K.P.N. (2012). Quest for the kondlo – Zululand's forgotten mystery bird. ShukerNature
http://www.karlshuker.blogspot.com/2012/02/quest-for-kondlo-zululands-forgotten.html 21 February. Accessed 29 January 2013.

If no author is given, simply begin the reference with Anon. and then give the article's publication year, title, etc as above.

References inserted directly in the paper's main text should take the form of: Smith (1988), Jones (1989). Or, if cited within brackets in the main text, they should take the form of (Smith, 1988; Jones, 1989).

Spelling/Punctuation Miscellanea

Always utilise the UK-English spelling variant of a given word if it differs from the American-English equivalent, e.g. colour, not color; organise, not organize; through, not thru; grey, not gray.

In text, please insert only a single space, not two spaces, between the full stop (period) at the end of one sentence and the capital letter beginning the first word of the next sentence.

Measurements:

Wherever possible, the metrical system of measurements should always be employed in manuscripts. However, when directly quoting historical documents that contain imperial measurements, it is permitted to retain these for purposes of accurate reporting, provided that their metrical equivalents are included directly alongside them in square brackets, e.g. 1 ft [30 cm].

Lightning Source UK Ltd.
Milton Keynes UK
UKHW02f0847081217
314092UK00004B/385/P

9 781909 488496